Basic Principles of Plates and Slabs

for safer and more
cost effective structures

P. G. Lowe
FIPENZ, FIEAust
Department of Civil Engineering,
Universities of Auckland,
New Zealand and Sydney, Australia

Whittles Publishing

CRC PRESS

Published by
Whittles Publishing Ltd.,
Dunbeath Mains Cottages,
Dunbeath,
Caithness, KW6 6EY,
Scotland, UK
www.whittlespublishing.com

Distributed in North America by
CRC Press LLC,
Taylor and Francis Group,
6000 Broken Sound Parkway NW,
Suite 300,
Boca Raton, FL 33487, USA

ISBN 1-870325-44-3
US ISBN 0-8493-9642-5

Typeset by
Mizpah Publishing Services Private Limited, Chennai, India

Printed by
Bell and Bain Ltd., Glasgow

Notation

═══	Simply supported edge
× × ×	Fixed (clamped) edge
———	Free edge
( ) . ( )	Scalar product
$( ) \times ( )$	Vector product
A	Slab area
B	Boundary length, perimeter
D	Plate stiffness
E	Young's Modulus (the context will prevent confusion with next symbol)
E, F, G	Coefficients of first fundamental form
L, M, N	Coefficients of second fundamental form
i, **j**, **k**	Cartesian unit vectors
I, **J**	Polar unit vectors
k	Constant curvature
L	Span, beam or plate
M_p	Full plastic moment
m_p	Full plastic moment
$M_{\alpha\beta}$	Moment stress resultants
Q_α	Transverse-force stress resultants
n	Unit normal vector
p	pressure loading
p_L, p_c, p_u	Collapse pressures
r, θ	Polar coordinates
r	Radius (or position) vector
s	Arc length
$( )_s$	Derivative with respect to s
t	Plate thickness
u, v	Surface coordinates
$( )_u$, $( )_v$	Derivatives with respect to u, v
w	Transverse displacement
x, y, z	Cartesian coordinates
$(. .)_{,x}$	Derivative with respect to x
α	Parameter $= B^2/A$
β	Parameter $= \gamma/\alpha$
γ	Parameter $= p_c A/M_p$
κ	Curvature
κ	Curvature matrix
ν	Poisson's ratio
$( )_L$	Lower bound collapse value
$( )_c$	Exact collapse value
$( )_u$	Upper bound collapse value
$( )^{IV}$	Fourth derivative, with respect to x, or according to context. Also $( )^{I}$, $( )^{II}$, $( )^{III}$ used
δ_{ij}	Cartesian metric tensor
e_{ij}	Strain tensor
t_{ij}	Stress tensor

Ki toku hoa rangatira
kua haere kia rangi

Janet O. Lowe
1935–2003

Contents

View of the earliest experimental studies on the "preferred construction method", *c.* 1991, showing the assembly under test. See Chapter 6, Construction and Design, for the discussion of the hardware perfomance.

0. Preliminaries

0.0 Motivation

This is a book about floor structures in buildings. As an indication of the relative importance of floors compared with all other components, we note that frequently the majority of the load-bearing material in a building is incorporated in the floors. Obviously they are important parts of the structure. We begin with the theory of such structures: the floor is then referred to as a plate. Later we move on to consider construction and design aspects of these components. Then the floor is referred to as a slab. The space devoted to theory is considerably more than the construction/design content, but I personally regard these later construction-related sections of the book as the more important from the point of view of stimulating discussion of this very important subject. My aim is to stimulate reader interest for them to want to explore the subject further. There are still many important aspects of this age-old subject to be investigated in the light of modern-day conditions!

This is a revision of an earlier theory-only book. The content has now been extended to include a brief introduction to design and construction as well as theory. The title has been altered to reflect the nature of these changes. As we might expect, the subject has moved on in significant ways since the publication of the earlier book. My original intention had been to make changes just to take account of recent developments in theory. However events have overtaken this intention. The major of these are the attacks on the World Trade Center on 11th September 2001. Those events have many facets, some of which have technical and construction-related implications. There are many very important and as yet unanswered questions regarding key aspects of the planning and design decisions taken and the building technology employed in the construction of the Twin Towers. These technical repercussions have not yet run full course. Seen at this distance from the "9/11" events, the apparent reluctance of relevant parties to face some of the technical questions raised by the collapse of the Twin Towers is, at the very least, puzzling. But above all it was the long span floors in the towers that failed first and brought the entire construction down.

The revisions I have found myself almost driven to make to my book have been much influenced by this situation. The earlier work as I have already noted was confined purely to theory. In this revision I have introduced what I consider to be relevant discussion on design and construction as well. This is an unusual combination, but the post "9/11" world calls for us to consider unusual situations. It was the choice of floor structure type and how it was supported in the World Trade Center Twin Towers, and particularly vulnerability in fire, that was the Achilles heel of those buildings. Reports are saying many things including that the relevant fire tests were never carried out. My book is primarily considering floor structures. Hence it has seemed to me important to expand the scope of the earlier analysis-only book to include as much discussion of construction and design as reasonably could be incorporated in the revised work. The emphasis in this part of the revisions has been on new and unconventional technology, since the conventional solutions to practical floor design and construction problems can be read about elsewhere and such solutions were available to the Twin Towers' designers at the time of their construction.

0.1 The aim of the work

This is not a textbook in the conventional sense since a substantial part of the content is not at present included in current courses. The general topic of plate and slab structures seems to me to be a relatively neglected subject in the total scheme of instruction in many courses, so my aim is to discuss issues and I hope arouse interest in the subject. I am endeavouring to do this is by raising a number of matters for study that are not well known and yet hold some promise for further development. My premise is that improvements can be made in the cost, quality and lifetime structural performance of buildings and infrastructure, especially through improvements in the flooring systems employed. We need to have this confidence if other "Twin Tower collapse" situations are to be avoided in the future.

After describing some core theoretical topics in the main body of the book, a change of direction is made to more practical matters. A general discussion about aims and ideals leads to a description of a new and unconventional construction method for practical floor structures that has the very important feature of inherent, in-built, passive fire resistance. The hope of course is that this or some similar method of construction will lead to the improvements sought.

In the structural mechanics literature plates and slabs are dealt with much less frequently than is frame analysis even though in most buildings the floors account for the largest proportion of the structural material requirements. Theory and practice are usually kept separate in courses though much thought is sometimes given to integrating the parts into a whole educational experience. Thus actual construction practices are not commonly dealt with in structures-related courses. Usually theory precedes practical considerations of construction. Yet there are perfectly viable scenarios that could be implemented which reverse the order and consider the construction methods to be adopted and the end product, the floor in our case, before considering any of the theory or design aspects. Also we need to be clear that design is a quite different study to construction. Even if this assessment is only partially representative of the current situation, these broad trends may account for some of the major deficiencies showing up in some modern construction, with the Trade Center Tower extreme vulnerability again as the prototype situation to be avoided at all cost.

In this book the choice has been made to limit discussion to plates and slabs, and then to relate the theoretical content directly to issues of practical construction. The reader should therefore seek books where other structures-related topics can be studied co-laterally to give a balanced treatment of the whole field. The emphasis here is on simple mechanics and solutions of the models derived. Finite element methods, important though they are, are not dealt with. They provide a good example of the need for co-lateral study.

It is unusual to find several theoretical models of plates and slabs dealt with in a single book. Here three quite separate models of structural response are described. It is hoped that readers will have their interest aroused by these brief windows into different areas of a large subject. After dealing with elastic theory we then devote rather more space to ultimate load analyses of these vital structures. Emphasis is placed on methods capable of producing safe lower bounds on the collapse loads. The better known alternatives can generally only be used to produce unsafe upper bounds. A name, the Comparison Method, has been coined for this lower bound method. The chapters dealing with theory conclude with a brief treatment of an optimal model for plate bending mechanics.

As already noted, the later parts of this book are devoted to discussion of some of the issues relating to design and particularly to construction. Construction does not fit neatly with theory or computation but it is a vital topic and deserves more attention in the education of future professionals.

0.2 Acknowledgements

It is a very pleasant duty to acknowledge those who have influenced, helped and encouraged me over the years. First there were the teachers and supervisors and later the colleagues. From my school days one teacher stands out, the late W. A. C. Smith. It was he who first aroused my interest in mathematics. Then the late Professor Cecil Segedin continued this process of interest stimulation very effectively for me in the university environment, and encouraged me in later years. During periods of study and employment in Sydney, London, Cambridge, Glasgow and Auckland I have had the good fortune to study or work with many colleagues and friends, including the late Bill Wittrick, Ronald Tiffen, the late Ronald Jenkins, Jacques Heyman, Ken Livesley, Christopher Calladine, Andrew Palmer, Robert Melchers, Ian Collins, John Butterworth, Ray Thompson, Rex Williams, Walter

Schumann, John Allen, Rob Irwin, Paul Wymer, Mark Batchelar, John Ingram, and others. They have discussed and helped clarify many topics with me. Last but not least the students in my classes, particularly the postgraduates, including Choong Kay Cheong, Steven Dwyer, Nicholas Charman, Anthony Van Erp, Samir Kanji and Tarek Omar, who have added important ingredients to my thoughts on parts of the subject. I thank them all.

There are three special tributes I wish to pay: to Robert Dykes (1927–1980), a friend and colleague at Strathclyde who died far too young; to my long-time friend and professional colleague Frank Hodgson (1926–2002), architect turned engineer and man of many parts, and to Catherine Logan (née Drummond) (1942–2003) who provided me with important assistance, typed the original manuscript for this book and kept me informed about Glasgow-related matters in the years from then till now. My thanks also go to Dr Keith Whittles and his staff; I wish him and his publishing house every success.

Finally it is to my late wife, Janet, that I owe the most. Though she did not share my interests in technical matters she helped me maintain momentum through both bright and dark times. It is to her that this work is dedicated.

Peter Lowe
Woollahra
NSW, Australia
May 2005

1 Review—the change from one to two dimensions

1.0 Introduction

At various points in the later chapters, but especially in connection with the study of the relevant geometrical content of the subject, it will be convenient to offer a vector formulation. It is proposed therefore to review those aspects of vectors which will be drawn upon later.

1.1 Vectors—algebra

A vector is a quantity which possesses both *magnitude* and *direction*. Examples of vectors are forces and surface slopes. There is an important additional class of vectors with which the geometrical part of the subject will deal— the class of *position vectors*.

Visualize a coordinate system, say a rectangular cartesian system of straight, orthogonal axes (Fig. 1.0). Any point such as **P**(x, y, z) when referred to the origin **O**, for example, will be spoken of as possessing a *position*, or *radius*, vector **r** as shown in the figure.

Once the origin has been chosen and the point **P** selected then the position vector **OP** will be fixed, and will have the attributes of a vector, namely of magnitude and direction.

When a vector has the special quality that its magnitude is fixed and equal to unity, then that vector will be called a *unit vector*. This is the defining quality; a unit vector has unit magnitude. There is no restriction placed on the direction of the unit vector.

Associated with the rectangular cartesian axes x, y, z (Fig. 1.0), it is conventional to define three unit vectors **i**, **j**, **k**, directed in turn along the axes x, y and z. Thus **i** is by definition the unit vector directed along the x axis, **j** along the y and **k** along the z axis. In this case, these three special unit vectors *do* have restricted directions—those associated with x, y and z respectively.

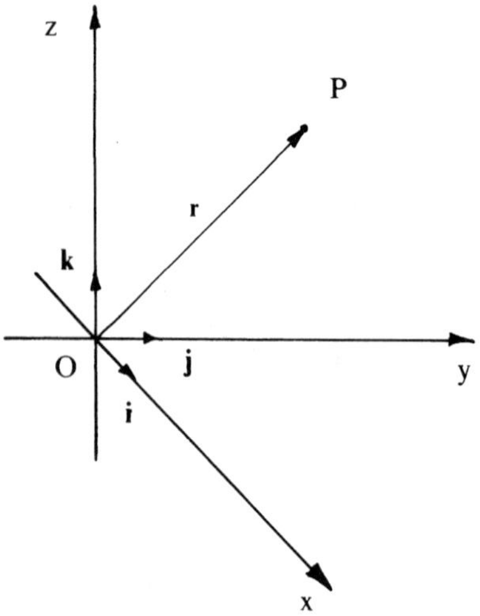

Figure 1.0 The position vector.

The position vector **r** of point P, with respect to O, can be written as

$$\mathbf{r} = x\mathbf{i} + y\mathbf{j} + z\mathbf{k} \tag{1.1.0}$$

where now x, y, z are the *coordinates* of P. Alternatively, x, y and z can be called the *components* of **r** with respect to the axes x, y, z. They are the lengths of the sides of the rectangular parallelapiped with OP as the diagonal.

Vectors may be added and subtracted, according to the parallelogram rule. If two forces acting at a common point are added together, a *single* equivalent force, often known as the *resultant-force*, is produced. The resultant is the *diagonal* of the parallelogram of which the original two forces are adjacent sides (Fig. 1.1). This equivalent single force $\mathbf{F}_R$ is correct for magnitude and direction. Algebraically this addition is written

$$\begin{aligned}\mathbf{F}_R &= \mathbf{F}_1 + \mathbf{F}_2 \\ &= \mathbf{F}_2 + \mathbf{F}_1.\end{aligned} \tag{1.1.1}$$

Subtraction is performed in a similar manner:

$$\mathbf{F}_2 = \mathbf{F}_R - \mathbf{F}_1$$

or (1.1.2)

$$\mathbf{F}_1 = \mathbf{F}_R - \mathbf{F}_2$$

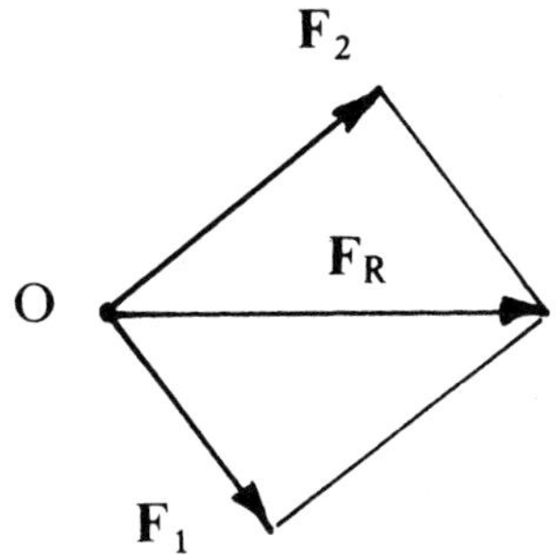

Figure 1.1 Vector addition.

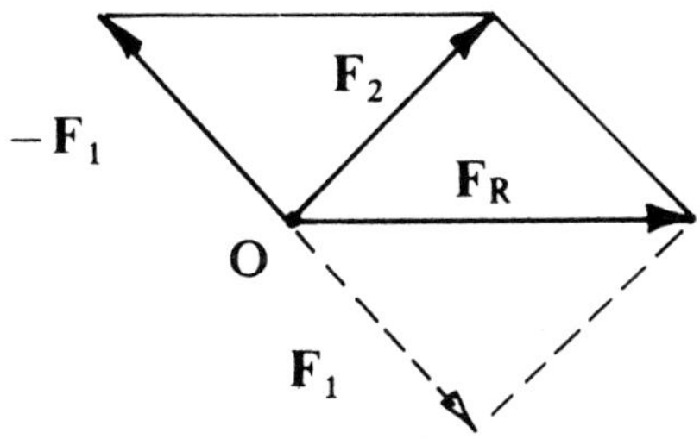

Figure 1.2 Vector subtraction.

and the subtraction can be illustrated on a diagram where it is seen that $\mathbf{F}_2$ results from the addition of $\mathbf{F}_R$ and $(-\mathbf{F}_1)$ (Fig. 1.2).

This discussion of addition and subtraction of vectors has been couched in terms of force vectors, but the same rules and physical meanings can be attached to these operations applied to other sorts of vectors. The only requirement is that the vectors be of the *same type* for them to be combined, namely they must both be forces or radius (position) vectors, for example.

The algebra of vectors can be extended beyond addition and subtraction to include multiplication, but now there are *two* types of multiplication. First there is the operation of multiplying two vectors together so as to form a scalar—this is called *scalar multiplication*. Recall, too, that a scalar is an ordinary number, which has magnitude but *no* directional quality associated.

This operation of *scalar multiplication* of two vectors **a** and **b** is defined to be

$$\mathbf{a} \,.\, \mathbf{b} = |\mathbf{a}|\,|\mathbf{b}|\cos\theta \tag{1.1.3}$$

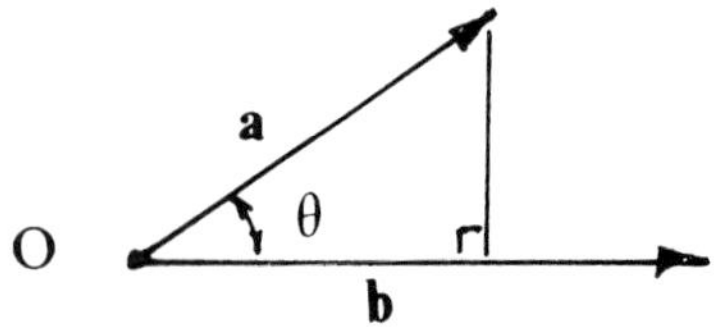

Figure 1.3 Scalar product.

where |**a**| is the symbol used to indicate *magnitude* of **a** and θ is the angle between **a** and **b** (see Fig. 1.3). From the nature of the right-hand side of (1.1.3) it can be confirmed that **a** . **b**, the scalar product of **a** and **b**, is indeed a scalar. The reason for introducing this product is one of convenience. An interpretation is that **a** . **b** is the scalar value of the magnitude of **b** multiplied by the component of **a** in the direction of **b**. This latter quantity is the magnitude of the vector **a** thought of as composed of two components, the one *along* **b** and the other normal to **b** (see Fig. 1.3).

Thus

$$\mathbf{a} \,.\, \mathbf{b} = b(a \cos \theta)$$

Alternatively, the roles of **a** and **b** can be exchanged and the result is unaffected. Hence **a** . **b** = **b** . **a**, where the scalar product is *associative*.

Return to consider the position vector **r** of (1.1.0). The most convenient manner for finding the components of **r**, given (1.1.0), is to form the scalar products **r** . **i**, **r** . **j**, **r** . **k** in turn. Then, because **j** and **k** are mutually perpendicular to **i**, their respective θ's are π/2 and hence

$$\mathbf{i} \,.\, \mathbf{j} = 0 = \mathbf{i} \,.\, \mathbf{k}$$

Hence

$$\mathbf{r} \,.\, \mathbf{i} = x\mathbf{i} \,.\, \mathbf{i} + y\mathbf{j} \,.\, \mathbf{i} + z\mathbf{k} \,.\, \mathbf{i} \qquad (1.1.4)$$
$$= x.$$

This result follows since **i** . **i** = 1, **i** being a unit vector and θ here is zero.

Relation (1.1.4) serves to illustrate a useful role of orthogonal unit vectors and the scalar product.

Exercise: The magnitude of a vector, **r**, can conveniently be found by noting that **r** . **r** is a scalar and has magnitude |**r**| . |**r**| cos θ = r^2, where |**r**| is the magnitude of **r**. Hence show that the magnitude of **r** with x, y, z components (3, 4, 5) is $5\sqrt{2}$.

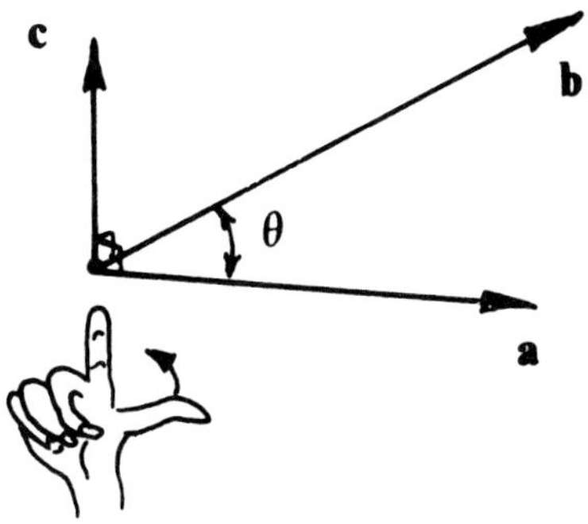

Figure 1.4 Vector product.

The second type of multiplication of vectors is the *vector product*, so called because the result of multiplying two vectors in this case is to produce a further vector. The operation is denoted by ( ) × ( ). Hence

$$\mathbf{a} \times \mathbf{b} = \mathbf{c}$$

where $|\mathbf{c}| = ab \sin \theta$ and has a direction given by a right-hand rule with respect to **a** and **b** (see Fig. 1.4).

Again, the motivation for introducing the vector product is one of utility.

Note now that if the vectors multiplied in a vector product are reversed, then the *direction* of the resulting vector is *reversed*. Hence the operation ( ) × ( ) is *not* commutative, that is the order of multiplication of terms cannot be reversed without changing the result.

Exercise: Consider the trio of fixed unit vectors **i, j, k**. Suppose x, y, z form a right-handed triad of axes. This means that applying the right-hand rule to carry x into y (by motion of the thumb) ensures that the forefinger is pointing along the positive z axis during the process.

Hence show that $\mathbf{i} \times \mathbf{j} = \mathbf{k}$, $\mathbf{k} \times \mathbf{j} = -\mathbf{i}$, $\mathbf{i} \times \mathbf{i} = 0$ and similar results.

1.2 Vectors—calculus

The discussion of vectors thus far has been concerned with *algebraic* operations on vectors. Note, too, that no operation of *division* has been defined.

There is use for a little of the calculus of vectors, particularly differentiation of vectors, in the following chapters.

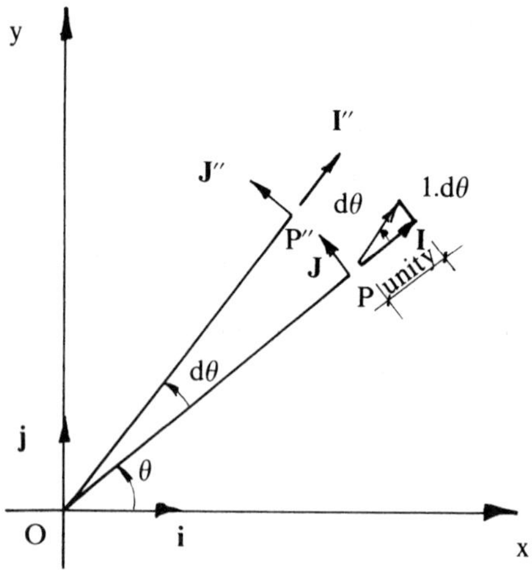

Figure 1.5 Polar coordinate—unit vectors.

A vector, such as a position vector (**r**) changes with choice of the point P. Hence **r** can be thought of as a function of the coordinates of P. As different points P are chosen, so **r** changes and hence rates of change and differentiation become natural developments.

Consider a particular circumstance of change. The polar coordinate system (Fig. 1.5) is specified by r, θ. Associated, and defined are the unit vectors **I**, **J**. As r and θ change then **I** and **J** are likely to change, but by how much?

Note first that the *magnitudes* of **I** and **J** do no change, because by definition, they are unit vectors, and hence their magnitudes remain at unity. However the *directions* of **I**, **J** may change, and hence, so far as unit vectors are concerned, changes are limited to changes of direction.

In the present case, if **r** changes to $\mathbf{r} + \mathrm{d}\mathbf{r}$ but θ is held constant then P moves to P′ where $\mathrm{PP}' = \mathrm{d}\mathbf{r}$. However, the directions of $\mathbf{I}'$ and $\mathbf{J}'$, the new unit vectors at P′ are unchanged, and hence $\mathbf{I}' = \mathbf{I}$, $\mathbf{J}' = \mathbf{J}$.

But should θ change to $\theta + \mathrm{d}\theta$ and P move to P″ (Fig. 1.5) then the directions of the associated **I**, **J** do change. From the geometry of the figure it is seen that

$$\mathbf{I}'' = \mathbf{I} + \mathrm{d}\theta \,.\, \mathbf{J}$$

and (1.2.0)

$$\mathbf{J}'' = \mathbf{J} - \mathrm{d}\theta \,.\, \mathbf{I}.$$

Adopting the notation of scalar differential calculus, then for $d\theta$ an infinitesimal, $\mathbf{I}'' = \mathbf{I} + d\mathbf{I}$ and

$$\frac{d\mathbf{I}}{d\theta} = \mathbf{J}, \quad \frac{d\mathbf{J}}{d\theta} = -\mathbf{I}. \tag{1.2.1}$$

Hence there arise differentials of vectors. In the use made of the concept of differentiation, the important role of unit vectors will emerge. In summary, any *change* in a unit vector must occur *normal* to the current direction of the unit vector, because the change to a unit vector is associated with change of direction, but not of magnitude.

1.3 Matrices

Some familiarity with simple matrix operations will be useful later. A matrix is a rectangular array of numbers, sometimes with equal numbers of rows and columns, which can be looked upon as an extension of the concept of a vector. Once the coordinate system to be adopted is specified, then a single list of three numbers (the components of a vector) specifies a vector. Thus the position vector $\mathbf{r}$ of earlier can be specified by

$$\mathbf{r} = (x, y, z) = \begin{bmatrix} x \\ y \\ z \end{bmatrix}.$$

Another notation is

$$r_i = \begin{bmatrix} x \\ y \\ z \end{bmatrix} \begin{matrix} 1 \\ 2 \\ 3 \end{matrix}$$

where the "free" subscript i can take on the value 1, 2 or 3 to identify the x, y or z component. Such a notation emphasises the *single subscript* aspect of a vector. Vectors, too, can have more than three components, though interest here is limited to three component vectors.

In contrast, matrices are *double subscript* objects, since two subscripts are needed to specify first the row and second the column which a particular element (or component) occupies. Of special interest are the simplest matrices which consist of just two rows and two columns and hence four elements or components at most.

Consider such an array

$$\mathbf{A} \equiv \begin{bmatrix} a_{11} & a_{12} \\ a_{21} & a_{22} \end{bmatrix} = a_{ij}. \tag{1.3.0}$$

Notation can be confusing. As far as possible, **bold** face upper case letters will be used to represent *matrices*, the size of the matrix (the number of rows/columns) being indicated either explicitly or from the context. In contrast, lower case bold face letters will be reserved for *vectors*.

Should it happen that a_{12} equals a_{21} in (1.2.2), then the matrix **A** is said to be symmetric. Most physical examples of matrices are symmetric matrices. Like vectors, matrices possess a non-commutative aspect in multiplication but they are associative in addition and subtraction. To be added or subtracted, matrices must be of similar shape—the same numbers of rows and columns—when the result of addition is that similarly positioned elements in the matrices are added to produce a new, similarly positioned element in the resulting matrix.

Matrix multiplication is achieved by taking a row in the first matrix and multiplying this, element by element, with a column of the second matrix, and summing the terms so found. The resulting new element is positioned in the result matrix in the position with the original row and column numbers of the row and column being multiplied.

It is necessary therefore that for two matrices to be multiplied they must be *conformable*, that is, have the same number of columns in the first as there are rows in the second. The resulting matrix will have a "size" given by the rows of the first and the columns of the second. Many important physical cases are concerned with square matrices and hence with equal numbers of rows and columns, when the above rules are much simplified.

Example: Given

$$\mathbf{A} = (2)\overset{(3)}{\begin{bmatrix} 1 & 2 & 3 \\ 4 & 5 & 6 \end{bmatrix}},$$

$$\mathbf{B} = (3)\overset{(2)}{\begin{bmatrix} 1 & 2 \\ 3 & 4 \\ 5 & 6 \end{bmatrix}},$$

Find $\mathbf{C} = \mathbf{A} \,.\, \mathbf{B} = c_{ij}\,.\ (i = 1, 2, j = 1,2)$.

Consider

$$c_{11} = \overbrace{1 \times \underbrace{1 + 2}_{} \times 3 + 3 \times 5}^{\text{first row of A}} = 22,$$

first column of B

$$c_{12} = 1\times2+2\times4+3\times6=28,$$
$$c_{21} = 4\times1+5\times3+6\times5 = 49,$$
$$c_{22} = 4\times2+5\times4+6\times6=64.$$

$$\therefore \mathbf{C} = \begin{bmatrix} 22 & 28 \\ 49 & 64 \end{bmatrix}$$

1.4 Statics—equilibrium

Newton's second law of motion for a particle of mass m states that the rate of change of momentum is proportional to the force acting on the particle, and, as this is a vector equation, the direction of the velocity increase (the acceleration) is the same as that of the force acting. The main interest in the study of plates to follow is of equilibrium. Hence there are no accelerations, and as it will be assumed that the body was initially at rest, it therefore remains at rest, the resultant force acting on any particle must be zero, and the problem is one of statics.

Particles as such are seldom encountered in structural mechanics, and the concept of a body larger than a particle is therefore needed. Then the relevant law of statics is that the *resultant force* on this body must be zero, and the *resultant moment* of all the forces acting on the body about an arbitrary point must also be zero.

Generally, the moment of a given force about a chosen point changes with the choice of point. However, statics is concerned with a collection of forces acting on a body, and this system of forces has a zero resultant because the body is at rest. As the result of this, the resultant moment, made up of the vector addition of the moment of each individual force about the chosen point, if zero for *one* point will be zero for *all* points. Expressed in slightly different terms, if a system of forces has the property that the resultant moment of these forces is the *same* for all choices of point about which moments are taken, then the force system is equivalent to a *couple*—that is two equal and opposite forces not acting in the same line. If, in addition, the moment so calculated is zero, then the force system is equivalent to two equal and opposite forces but now acting collinearly. Such a force system is an *equilibrium* system. The words "statics" and "equilibrium" are used as alternatives but should really be used together to describe *static equilibrium*—the system being at rest in equilibrium.

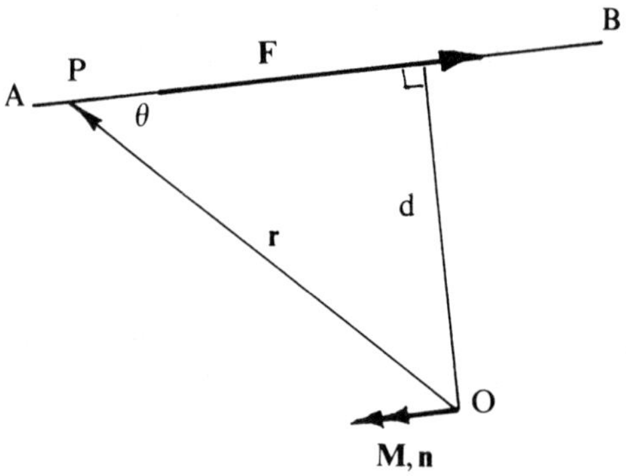

Figure 1.6 Moment of a force.

The general definition of the moment (**M**) of a force (**F**) about a point O is that

$$\mathbf{M} = \mathbf{r} \times \mathbf{F} \tag{1.4.0}$$

where **r** is *any* radius vector joining O to a point on the line of action of **F**. From the definition of vector product, **M** is a vector, normal to the plane of **r** and **F**. The moment is the *turning effect* of **F** about O.

All the examples presently of interest are two-dimensional in the sense that all **F** and **r** lie in a common plane. Hence all the **M** vectors will be *parallel* and can be treated as scalars, to be added or subtracted. From Fig. 1.6 it is seen that

$$\begin{aligned} \mathbf{M} &= \mathbf{r} \times \mathbf{F} \\ &= (\mathrm{r} \,.\, \mathrm{F} \,.\, \sin\theta)\,\mathbf{n} \qquad (1.4.1) \\ &= (\mathrm{F} \,.\, \mathrm{d})\mathbf{n} \end{aligned}$$

since $\mathrm{d} = \mathrm{r} \sin\theta.$

Here **n** is a unit vector normal to the plane of **r** and **F**. If all the **r** and **F** vectors lie in one plane, then all the **M** vectors will be parallel to **n**, and can be simply added or substracted, with **n** as a common factor.

The conditions for static equilibrium are then that

$$\mathbf{F}_\mathrm{R} = 0, \quad \mathbf{M}_\mathrm{R} = 0 \tag{1.4.2}$$

where $\mathbf{F}_\mathrm{R}$ and $\mathbf{M}_\mathrm{R}$ are, respectively, the *resultant force* and *resultant moment* of the individual forces about an arbitrary (i.e. any) point.

Exercise: Assume a right-handed cartesian coordinate system and associated unit vectors **i**, **j**, **k**. A force **F** with components (1, 1, 1) acts through the point, Q,

(0, 1, 0). Show that the moment (**M**) of **F** about point P with coordinates (1, 0, 0) is given by a vector with components (1, 1, −2). Show too that the axis of the moment **M** passes through R(2, 1, −2) and that PR is normal to the plane of **F** and P. Check that $M_p = \sqrt{6}, PR = \sqrt{6}, QR = \sqrt{8}$.

1.5 Summation convention and index notation

When multiplying vectors or matrices, many number pairs are generated and summed to find the new elements. If it is convenient, the *summation convention* may be used from time to time.

A vector **a** is described by the components a_x, a_y, a_z for example when a cartesian coordinate system is chosen. These three components can conveniently be referred to collectively as a_i (i = 1, 2, 3 or x, y, z). Here the subscript i is spoken of as a *free index*.

No free index denotes a scalar, *one* free index denotes a vector, *two* free indices denotes a matrix. For example, a_{ij} where both i(= 1, 2, 3) and j(= 1, 2, 3) are free indices, denotes an element in a matrix.

The scalar product of two vectors **a**, **b** can be conveniently represented in index terms as

$$\mathbf{a} \,.\, \mathbf{b} = a_i \,.\, b_i = a_1b_1 + a_2b_2 + a_3b_3.$$

Here the *repeated* index i can be seen to imply the summation shown, of terms of similar type added together. This is the *summation convention*, where the repeated index $a_i \,.\, b_i$ implies the sum of terms obtained by letting i take on the values 1, 2, 3 in turn.

Exercise: Two-by-two matrices A, B can be denoted by a_{ij}, b_{mn} where i, j; m, n can take the values 1, 2. Show that if C = AB then in index form $C_{ij} = a_{im}b_{mj} = a_{i1}b_{1j} + a_{i2}b_{2j}$, where now the repeated index m ensures the summation of row × column multiplication and summation.

Also, when convenient the following abbreviated notation will be adopted. A derivative such as dr/ds will be denoted by $(r)_{,s}$, or r' and a second or higher derivative likewise. Such notation is convenient typographically, but requires care by the reader in observing the placing of the comma (,). It is convenient also when used in conjunction with the summation convention. Some examples will be encountered later.

1.6 Beam bending

Simple beam calculations are often needed. In keeping with our main theme of plate and slab theory the following approach to beam bending leads into the plates and slabs.

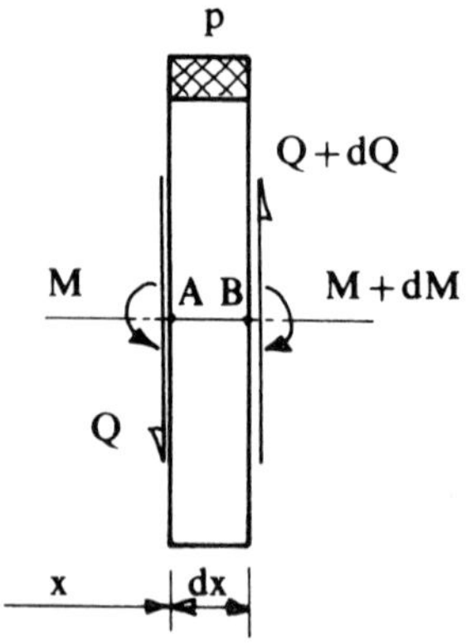

Figure 1.7 Beam element.

The primary dependent quantities are the bending moment M, shear force Q and transverse deflection w. Equilibrium is described by the gradient of the bending moment being equal to the shear force, and the gradient of the shear force being equal to the distributed load, p.

In symbols these equilibrium relationships are $dM/dx = Q$, $dQ/dx = p$, where x is the length coordinate describing the position on the beam, and p is the distributed load. Positive M causes upper surface tension, so hogging curvature, and positive Q turns the element counter-clockwise, see Figure 1.7. These are the same conventions as used in the later chapters. Usually the left-hand end is described by the coordinate $x = 0$.

First eliminate Q when

$$M'' = p$$

and the notation $(..)' = d(..)/dx$ has been adopted (1.6.0)

The integral of this equation is $\alpha x + \beta + px^2/2$, where α, β are unknown constants.

We shall discuss three types of beam response—elastic, plastic and optimal. They differ in the manner in which the material responds to being bent. The simplest is the plastic, usually termed *rigid plastic*, in which there is no visible bending of the beam until the moment reaches a limit, M_p, called the full plastic moment, and then a hinge occurs at the section where this moment has developed. Otherwise the member remains straight and this is described by the curvature being zero, $\kappa = w'' = 0$. Curvature (κ) is defined as the inverse of the radius of curvature. In practical beams the deflections must be kept small otherwise the beam will be too flexible for actual use. Hence slope of the beam (w') is small and the linear expression for the curvature ($\kappa = w''$) is of sufficient accuracy.

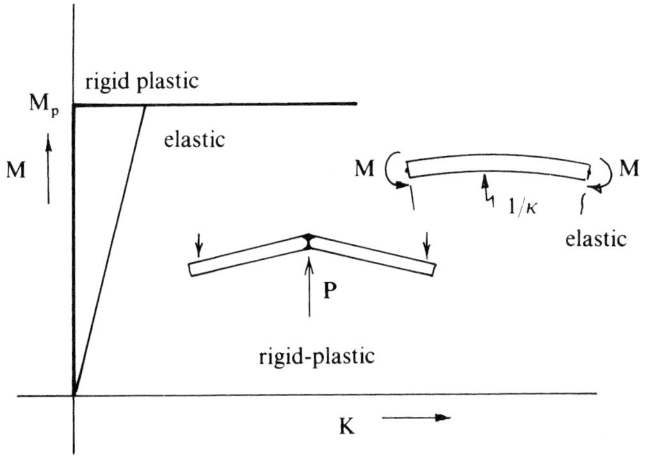

Figure 1.8 Moment-curvature relations.

The more familiar response is the *elastic bending* (Fig. 1.8) where curvature develops proportional to the moment divided by the stiffness, EI. Hence M = EI.w″. The third type is the *optimal* beam in which the curvature develops and remains constant, at k. So w″=|k|.

Logically the mechanics of rigid plastic beams is more simple than that of elastic beams, but traditionally is discussed after elastic theory. So we begin here with elastic beams.

Elastic beams

In symbols the elastic beam relationships are equilibrium described by (1.6.0) and the elastic property $EI.d^2w/dx^2 = M$.

Proceed by eliminating M to give the *elastic beam governing equation* for w as

$$EI\cdot\frac{d^4w}{dx^4} = EI.w^{IV} = p. \tag{1.6.1}$$

Here we have assumed the beam stiffness, EI, to be constant. This governing equation has a solution with four arbitrary constants A, B, C and D:

$$w = Ax^3 + Bx^2 + Cx + D + \frac{px^4}{(4!\,EI)}. \tag{1.6.2}$$

This is the general solution. The group of four polynomial terms with unknown coefficients is termed the *complementary function* (CF), to which must be added a *particular integral* (PI) for whatever external loads are applied to the beam. The 4! symbol denotes factorial 4, i.e. 4.3.2.1 = 24. The PI included in (1.6.2) is that for a constant uniformly distributed load, p/unit length. The reader should check by direct substitution of (1.6.2) into (1.6.1) that the governing equation is satisfied.

In some applications there may be other loads acting such as point loads (also called concentrated loads) which we shall denote by the symbol W. If such a load is located on the beam at $x = a$ then this will add a particular integral to the general solution for w of the form $W.z^3/3!EI$, where $z \equiv x - a$. The important point to note is that this particular integral is included only when z is **positive**, that is $x > a$. Remember that upper case W is a point (concentrated) load and must not be confused with lower case w which is the transverse displacement.

Consider an example of a simply supported beam, span L and loaded with a point load, W, at $x = a$, $0 < a < L$, that is at $z = 0$. Simple support here as for the plates and slabs later means the ends are supported so that $w = 0$ and $M = 0$ at each end. These are four conditions from which the four constants A, B, C and D are found.

Now $M = 0$ @ $x = 0$ gives $B = 0$ and $w = 0$ @ $x = 0$ requires that $D = 0$. At the other end, $x = L$, $M(L) = 0$ gives $w''(L) = 0$, where d( )/dx has been written as $(\ )'$. Thus $6AL + Wb/EI = 0$. Finally $w(L) = 0$ gives $AL^3 + CL + Wb^3/6EI = 0$. These two equations for A and C solve to give $A = -Wb/(6EI.L)$ and $CL = Wab(a+2b)/6EI$. Here $b \equiv L - a$, namely the distance from the right hand end of the beam to the point load, W, where a is the distance from the left-hand end.

Since we now know the four arbitrary constants, two of them being zero in this case, the deflection w(x) can be written down for any point on the beam, and the M(x), Q(x) values likewise can be written down.

Suppose we wish to know the maximum deflection, $w_{max,}$ and where along the beam it occurs. The deflection will be maximum where the slope $w'(x) = 0$. Now $w' = 3Ax^2 + C$, with no contribution from the W particular integral if we assume that $a > L/2$. We can always do this, if necessary, by viewing the beam from the other side.

Thus $w' = 0$ when $x^2 = -C/(3A) = a(a + 2b)/3$. The maximum value of w follows from substituting this value into the expression for w, namely

$$w_{max} = \frac{Wab\,(a+2b)}{9EI.L} \cdot \sqrt{\frac{(a(a+2b))}{3}}$$

Another quantity of interest is w(L/2), the mid-span deflection. This is given by $A(L/2)^3 + C(L/2)$. When simplified this becomes

$$w(L/2) = \frac{Wb\left(4a(a+2b) - L^2\right)}{48EI}\ .$$

As the load moves along the beam the maximum deflection always remains nearer the mid-span than the load itself. Only when the load is at mid-span does the load catch up the maximum displacement. The ratio of these two values, $w_{max}/w(L/2)$, is always > or = 1. It is left as an exercise for the reader to show this, and that the ratio never exceeds $16/(9.\sqrt{3})$, which is 1.0264. So for simply supported beams, loaded in any manner with gravity type loads, the maximum displacement can be safely approximated by increasing the mid-span displacement by 3%.

This result has been worked out for the simplest of loads, the single concentrated load. But the result holds true for any distribution of load so long as it is of gravity type, namely a downward load: also excluded are applied moments.

Plastic beams

Now we come to the plastic beam. It is usual to neglect the elastic deflections as being small compared to the plastic deflections that occur as a sufficient number of hinges develop to allow a mechanism to form and the beam to collapse. The beam segments between hinges remain straight and are then usually referred to as the rigid plastic response. Then the simple beam above, if loaded beyond the elastic and into the plastic regime, will collapse when the moment at $x = a$ reaches the full plastic limit, M_P. The maximum moment occurs under the load, therefore

$$W_C.ab/L = M_P \text{ and so the collapse load, } W_C = M_P.\ L/(ab).$$

The shape of this plastically deforming beam follows from the solution of $\kappa = 0$ everywhere, except at the hinge. Thus $\kappa = w'' = 0$ integrates to $w = \alpha x + \beta - \phi.z$ where the z term is the particular integral describing the hinge rotation of amount ϕ at $z = 0$. In the present case the boundary conditions to find all the unknowns are given by $w(0) = w(L) = 0$ when $\beta = 0$ and $\alpha L - \phi b = 0$. The hinge rotation ϕ is independent of the collapse load and in a practical case will develop and increase the longer the load remains at the collapse value. The deflection under the load follows as $w(a) = \alpha a = \phi ab/L$.

Optimal beams

The third type of beam is an optimal beam where the beam section is arranged such that the maximum stress induced by the load is the same, constant value at every section. One way to achieve this is in a beam of constant depth with the strength provided by strong fibres such as carbon or steel in the outer faces of the beam and proportioned to mirror the bending moment. Then the beam curvature will be the same at all points except that the curvature will be sagging in some parts and hogging in others. The theory says that these parts must merge smoothly.

Such a beam is described by $w'' = \pm k$, where k is a positive constant. The 'sign' of M must be the same as the 'sign' of the curvature to ensure compatibility of moment and curvature. Indeed the geometry implied by the curvature determines all the features of the shape and moment distribution in the beam, which turns out to be determinate. As always equilibrium is described by $M'' = p$. The PI most needed is that describing the change from a hogging to a sagging curvature segment and is achieved by a term of the type $2k(z^2/2!)$. This makes the necessary adjustment by changing the curvature from what it was to an equal value of opposite sign and with a smooth, slope-continuous transition.

As an example consider a propped cantilever, span L and a uniformly distributed load, p/unit length, along the whole length of the beam. The optimal curvature condition, $w'' = -k$, integrates to $w = \Omega x + \theta - kx^2/2 + 2kz^2/2$. Here we are thinking of $x = 0$ at the left end being the simple support for the propped cantilever, and hence a sagging moment and curvature there. The z coordinate is at present undefined and corresponds to the point in the beam where the curvature changes to hogging with the fixed end of the cantilever at $x = L$. Thus there are three unknowns: Ω, θ and z. The three conditions are $w(0) = 0$, $w(L) = 0$ and $w'(L) = 0$. Hence $\theta = 0$, $\Omega L - kL^2/2 + kz^2 = 0$ and $\Omega - kL + 2kz = 0$.

Eliminate Ω and solve for z, when $z^2 - 2Lz + 0.5L^2 = 0$. Thus $z/L = (2 - \sqrt{2})/2 = 0.293$. This is z(L) and hence the z origin is at $x = L(1 - 0.293) = 0.707L$. The bending moment, $M = ax + b + px^2/2$ for a uniform load over the whole beam is the usual integral of (1.6.0). There are two unknowns, a and b. The boundary conditions are $M(0) = 0$, so $b = 0$; $M(0.707L) = 0$, i.e. at $z = 0$, when $0.707La + p(0.707L)^2/2 = 0$. Hence $a = -p \times 0.353L$. Then the bending moment and shear at all points can be written down. The bending moment is negative until after $x = 0.707L$, then it is positive. The same is true of the curvature, confirming that the bending moment and curvature are always of the same sign. All three of these response types—elastic, plastic and optimal, will be further developed in the context of plates and slabs.

Virtual work method

The approach used above to solve beam bending examples has been to integrate the governing equation. An independent and powerful alternative method uses virtual work as the means to solve similar problems. Consider the example of the beam collapsing plastically as in Figure 1.9. The central displacement shown as Δ cannot be calculated by our present means since it will continue to increase the longer the load is left on the structure in the collapse state. We could however calculate the displacement just as the hinge forms, and we can use Δ in our virtual work calculation.

The lower part of the figure can be thought of as a possible, virtual displacement and we then proceed to calculate and equate the internal and external work implied by this choice of virtual displacement pattern. The internal work is just

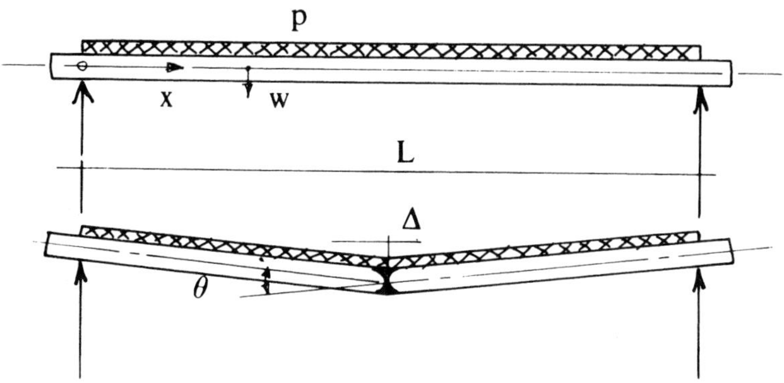

Figure 1.9 Beam —plastic collapse.

the product of full plastic moment at the centre, and the rotation of the beam segments there, $4\Delta/L$. The external work is the integral of all the pieces of distributed load multiplied by the virtual displacement of each portion of the load. In this case the internal work is twice the integral over half the length of $p(2\Delta x/L).dx$. This integrates to $pL\Delta/2$ and is equated to $4\Delta M_p/L$. The result is that the collapse value for the uniformly distributed load p_c is $8M_p/L^2$.

Similar but more complex calculations can be used to solve the same range of problems that the integration method can handle, namely small deflection and bending of the beam type needed.

Both the above methods produce exact solutions. Approximate solutions may be quite adequate in some applications. One method for producing approximate solutions is discussed in Chapter 3 under the finite difference method.

Exercises

(1) Consider a beam, single span L, and restrained against rotation at the ends. Let the loading be a concentrated load, W, placed at distance $a > L/2$ along the beam. In the elastic regime for the simply supported span we have discussed the maxima of displacement etc. above. So consider the plastic collapse conditions. Show that the collapse load is doubled by fixing the ends to $2M_p.L/ab$. Then calculate the displacement under the load when the beam is on the point of collapse after deflecting first purely elastically then with hinges forming at two sites in the beam and rotating there as the load increases and prior to the moment at the origin reaching the full plastic value and the beam collapsing. Show this deflection, $w(a) = \Delta$ to be $6EI\Delta = a^2 M_p$.

(2) Consider a uniform, continuous, elastic beam, simply supported at the ends and at the centre, of two equal spans L, with a hinge at a point in the left span at $x = \alpha L$, $\alpha < 1$. The loading of uniformly distributed type p kN/unit length extends over the whole span. We want to choose values for α, that is position the hinge, to achieve particular goals. The general shape of the beam is given by (1.6.2) with the addition of two further terms: these are a term θz_1, $z_1 = x - \alpha L$ to describe the rotation at the hinge and also a concentrated force term $- Rz_2^3/3!EI$, $z_2 = x - L$, tailored to describe the upward unknown reaction, R, at the centre support.

First note that the beam is determinate. This means that the internal forces and moments can be found from statics alone. There are good physical reasons why we might wish the beam to be determinate. For example, if one support settles a small amount no changes occur in the reactions if the structure is determinate. Without the hinge the beam is indeterminate and support settlement will cause changes in the reactions and so on. For all values of α other than $\alpha = 1$, the structure is not symmetrical: when $\alpha = 1$ we have two simple beams. We shall limit the span hinge to be in the left span in which case $\alpha<1$. We wish to consider two questions. First find the value for α to minimize the maximum bending moment, of either sign, in the beam. Then consider where to place the hinge if the need is to minimize the maximum displacement.

It is evident that the bending moment is symmetrical about the centre support for all values of α. But the deflected shape in general is not symmetrical. The maximum bending moment will be minimized when the sagging maximum is equal to the hogging maximum. Now the sagging maximum occurs at $x = \alpha L/2$ as can be seen from the simple span from $x = 0$ to $x = \alpha L$: the hogging maximum is always over the centre support. The end reactions are each $p\alpha L/2$ and the moment condition becomes

$$\frac{p(\alpha L)^2}{8} = -\left(\frac{p\alpha L}{2}\right)L + \frac{pL^2}{2}$$

This quadratic in α has one useful root of $2(\sqrt{2}-1) = 0.828$. Using this value in either the left or right hand side of the above expression gives the maximum moment to be $0.08579pL^2$. Show that the maximum displacement is then $0.00766pL^4/EI$ and that this occurs at x/L a little more than 1.5.

The deflected shape for minimum maximum moment is not symmetrical: there is rotation at the $x = \alpha L$ hinge of $0.031557pL^3/EI$ radians. The general expression for this hinge rotation, θ, is $(4 - 3/\alpha)pL^3/(12EI)$. To minimize the maximum displacement we must ensure that the maximum displacement in the left span is the same as that in the right. This can only be achieved if there is *no* rotation, θ at the span hinge and this requires that $\alpha = 0.75$. The moment over the centre support then rises to $0.125pL^2$ and the maximum displacement reduces to $w_{max} = 0.005416pL^4/EI$, at $x = 0.42L$.

In this final case both the bending moment and deflected shapes are symmetrical about the centre support line even though the structure clearly is not. Very rarely do the maxima of bending moment and displacement occur at the same point in a beam except where symmetry dictates they should be.

(3) Further, consider the propped cantilever optimum beam under uniform loading p, as on p. 20. Show that the moment volume V_M, defined as the integral over the length of the beam of the modulus of the bending moment, is given by $pL^3/6(2+\sqrt{2}) = 0.0488pL^3$. This is a measure of the minimum amount of moment resisting material capable of supporting the uniform loading p over the propped beam span L.

1.7 Conclusions

In this chapter a number of topics have been introduced which are essentially preliminary to a discussion of plate theory proper. Not all the topics need be mastered initially—for example the vector manipulations. However, later, when these topics can be used with some facility, they will allow the reader a fuller exploration of the whole subject of plate theories.

2 Statics and kinematics of plate bending

2.0 Introduction

In the Appendix the geometry of surfaces is developed. Relevant results will be used here for purposes of developing a suitable plate theory. The themes are the *statics* of a bent plate and the *kinematics* of plate bending. As remarked earlier, plates are essentially two-dimensional beam-type structures. In particular, the main interest is in plates which span transversely (horizontally say) and are loaded laterally (vertically say). Such a configuration produces a *bending* system in the plate.

For the present, all plates considered will be assumed to be homogeneous. Then the middle plane, that plane which is equidistant from the two outer faces of the plate, will remain *unstretched* by the bending and twisting actions. Parts of the outer layers will be compressed, extended (and sheared by twisting actions), but though the middle surface will bend into some surface shape it will not be stretched or sheared. The theories developed are all *small deflection theories*. This is equivalent to the geometrical requirement that the surface slopes are always small, that is, much less than unity.

2.1 The stress resultants

The plates of interest are those subjected to bending arising from transverse loading on the plate. There is a gradient of stress through the plate, but for purposes of formulating a working theory, stresses as such will not be used. Instead, certain important *stress resultants* will be defined. The same situation exists in beam theory where shear forces and moments are used. On a cut face in the plate material, x = const. (Fig. 2.0) let the stresses be referred to a right-handed cartesian system of axes (x, y, z). Than σ_{xy} is a shear stress, and acts on a face x = const. and in the y-direction, whereas σ_{xx} is a direct tension or compression

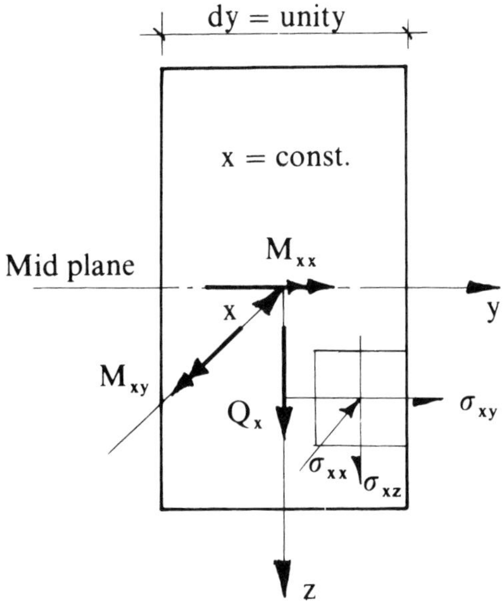

Figure 2.0 Stress and stress resultants, x = const.

(see Fig. 2.0). Since the section is being bent, σ_{xx} is likely to be compressive on one side of the mid-plane and tensile on the other. The same may also be true of the shear components. "Stress resultant" is the general term for some integral relation of stress through the thickness. Thus a typical stress resultant is

$$N_{xx} \equiv \int_{-t/2}^{t/2} \sigma_{xx} \, . \, dz. \qquad (2.1.0)$$

This is the overall compressive force on the segment shown and in the present case, this force will be zero, because the plate is being bent rather than compressed. However, the quantity

$$M_{xx} \equiv \int_{-t/2}^{t/2} \sigma_{xx} z \, . \, dz \qquad (2.1.1)$$

is in general not zero, and is a quantity of primary interest. This is the bending moment per unit length seen in plan and can be represented by the (double-headed) moment vector of turning *about* an axis parallel to the y axis (see Fig. 2.0).

The remaining quantities of interest are the twisting moment

$$M_{xy} \equiv \int_{-t/2}^{t/2} \sigma_{xy} z \, . \, dz \tag{2.1.2}$$

and the transverse shear force

$$Q_x \equiv \int_{-t/2}^{t/2} \sigma_{xz} \, . \, dz. \tag{2.1.3}$$

Each is also measured per unit length seen in plan.

The quantity

$$N_{xy} \equiv \int_{-t/2}^{t/2} \sigma_{xy} \, . \, dz \tag{2.1.4}$$

is, like N_{xx}, a membrane force, here a shear force, and these forces are not generated by transverse bending. Hence $N_{xy} = 0$ will be the situation for plates in bending.

The vector representation for M_{xy} (the twisting moment) is a turning action about the x direction across a section x = const., as indicated in Fig. 2.0, and Q_x is a conventional shear force vector directed along the z direction. The M_{xy} can be thought of as generated by a distribution of Q-type force of amount M_{xy}; this point will be raised again later.

Thus far the stress resultants associated with the cut face x = const. have been illustrated. Assuming still a right-handed cartesian coordinate system, then a cut on face y = const. must also be considered, when it will be seen (Fig. 2.1) that the relevant stress resultants are M_{yy}, M_{yx} and Q_y. By comparison with the quantities on the face x = const., there follow the definitions

$$\begin{aligned} M_{yy} &\equiv \int_{-t/2}^{t/2} \sigma_{yy} z \, . \, dz, \\ M_{yx} &\equiv \int_{-t/2}^{t/2} \sigma_{yx} z \, . \, dz, \\ Q_y &\equiv \int_{-t/2}^{t/2} \sigma_{yz} \, . \, dz. \end{aligned} \tag{2.1.5}$$

Note however that $\sigma_{xy} = \sigma_{yx}$ is an equilibrium requirement for the small element of material and hence

$$M_{xy} = M_{yx}.$$

To summarize then, the state of stress at a point in a plate will be represented by a group of stress resultants, the bending moments M_{xx}, M_{yy}; the twisting moments

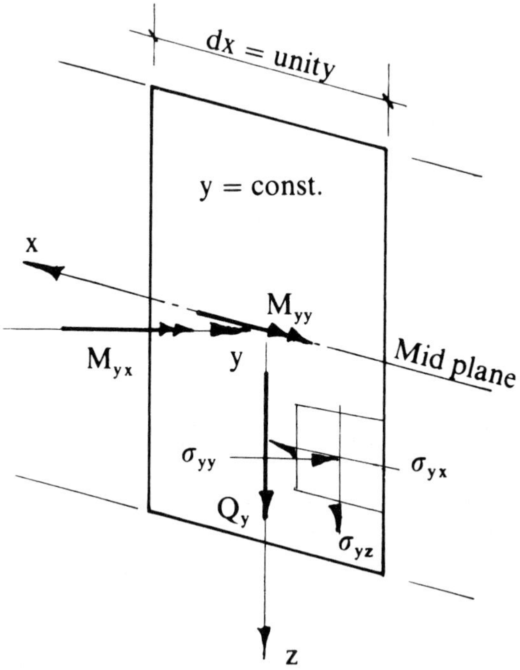

Figure 2.1 Stress and stress resultants, y = const.

$M_{xy} = M_{yx}$ and the transverse shear forces Q_x, Q_y. All these quantities are measured per unit length of the mid-surface. Hence, seen in vector terms, the cut faces of a piece of plate will be subjected to internal forces and moments as indicated in Fig. 2.2. The implied sign convention is then that *hogging* bending moments are *positive*.

The collection of moments M_{xx}, $M_{xy} = M_{yx}$, M_{yy} can conveniently be written as a 2 × 2 matrix, the moment matrix, thus

$$\mathbf{M} \equiv \begin{bmatrix} M_{xx} & M_{xy} \\ M_{xy} & M_{yy} \end{bmatrix}. \tag{2.1.6}$$

This is a symmetric matrix and once the three independent components of this matrix are known at a point for one choice of axes, then the values of the bending and twisting moment on any other transverse plane through this point, different from the x = const. and y = const. planes, can be found. Just how this can be achieved will now be considered.

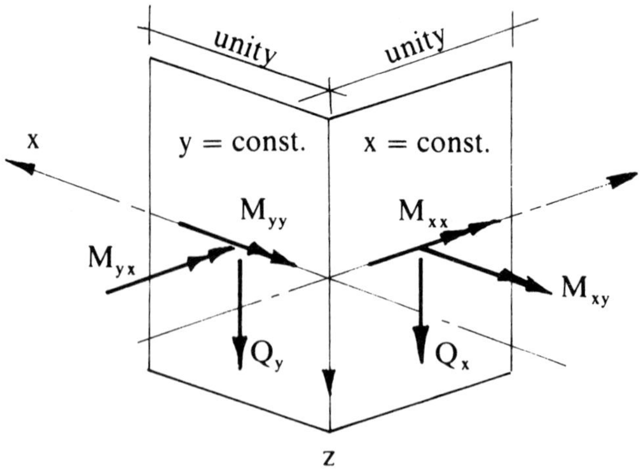

Figure 2.2 Stress resultants.

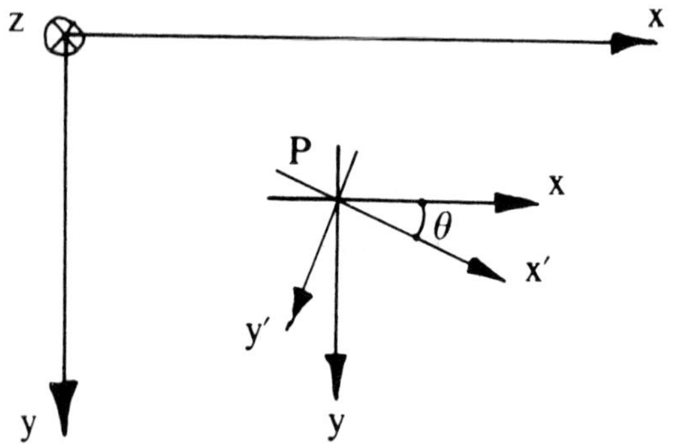

Figure 2.3 Rotation of axes.

2.2 Principal values

The moment matrix, **M**, when referred to x and y axes at a point P, is written

$$\mathbf{M} = \begin{bmatrix} M_{xx} & M_{xy} \\ M_{xy} & M_{yy} \end{bmatrix}. \tag{2.2.0}$$

Suppose it is desired to refer the matrix **M** to the choice of axes x′y′ at P (Fig. 2.3). What can be said about the values of the elements in the corresponding matrix **M′**? One way of proceeding is to consider the equilibrium of a small triangular element notionally cut from the plate and containing P (Fig. 2.4*a,b*).

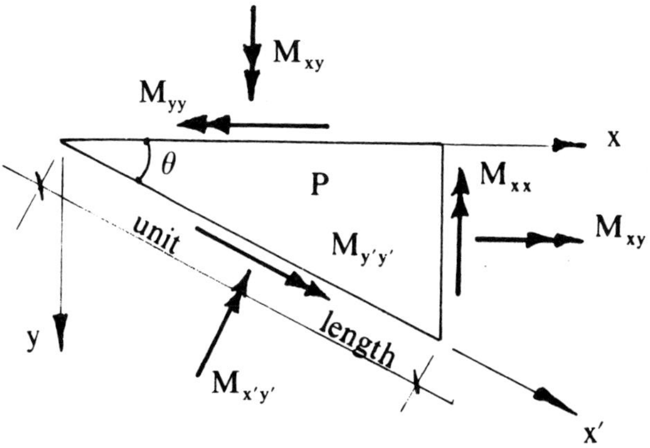

Figure 2.4(*a*) Change of axes I.

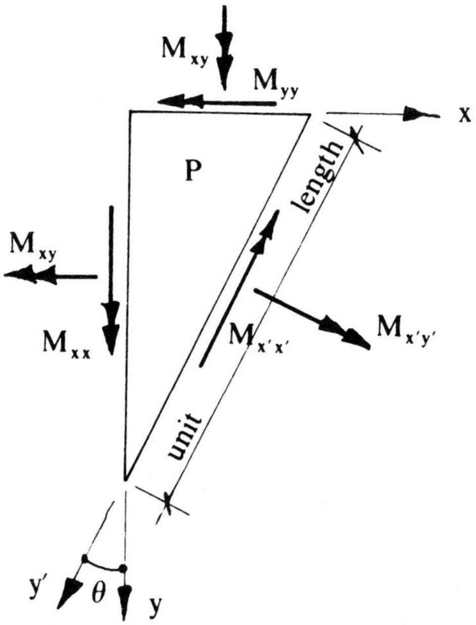

Figure 2.4(*b*) Change of axes II.

Recall first that all the quantities are computed per unit length, then resolve along, and normal to, x′ when

$$M_{y'y'} = M_{yy} \cos^2\theta + M_{xx} \sin^2\theta - 2M_{xy} \sin\theta \cos\theta$$
$$M_{x'y'} = (M_{yy} - M_{xx}) \sin\theta \cos\theta + M_{xy} (\cos^2\theta - \sin^2\theta). \tag{2.2.1}$$

Again, resolve along and normal to y′ direction (Fig. 2.4*b*), then

$$M_{x'x'} = M_{xx} \cos^2\theta + M_{xy} \sin^2\theta + 2M_{xy} \sin\theta \cos\theta$$
$$M_{x'y'} = (M_{yy} - M_{xx}) \sin\theta \cos\theta + M_{xy} (\cos^2\theta - \sin^2\theta). \tag{2.2.2}$$

It is worth pausing to consider the directions shown for quantities in Fig. 2.4*a,b*. By construction, all the edges of the triangular elements are along coordinate lines. The basic directions of quantities have been carried forward from Fig. 2.1 and 2.2, but where the face concerned bounds the far rather than the near side of this element, they by Newton's Third Law of Action-Reaction, reversed quantities are introduced. Such considerations account for the change in the M_{xx} direction in Fig. 2.4*a*, as compared with Fig. 2.2.

From (2.2.1), (2.2.2), given **M** and θ, the elements of **M**′ can be evaluated. But consider the coordinate change x, y → x′, y′. The change is one of rotation where

$$\begin{bmatrix} x' \\ y' \end{bmatrix} = \begin{bmatrix} \cos\theta & \sin\theta \\ -\sin\theta & \cos\theta \end{bmatrix} \begin{bmatrix} x \\ y \end{bmatrix} \equiv \mathbf{R} \begin{bmatrix} x \\ y \end{bmatrix}, \tag{2.2.3}$$

and the rotation matrix **R** is such that

$$\mathbf{R}^{I} = \mathbf{R}^{T} = \begin{bmatrix} \cos\theta & -\sin\theta \\ \sin\theta & \cos\theta \end{bmatrix}.$$

Now
$$\mathbf{RMR}^{T} = \begin{bmatrix} C & +S \\ -S & C \end{bmatrix} \begin{bmatrix} M_{xx} & +M_{xy} \\ M_{xy} & +M_{yy} \end{bmatrix} \begin{bmatrix} C & -S \\ +S & C \end{bmatrix},$$

where $C = \cos\theta$, $S = \sin\theta$.

Thus

$$\mathbf{RMR}^{T} = \begin{bmatrix} C & +S \\ -S & C \end{bmatrix} \begin{bmatrix} M_{xx}C + M_{xy}S & -M_{xx}S + M_{xy}C \\ M_{xy}C + M_{yy}S & -M_{xy}S + M_{yy}C \end{bmatrix}$$

$$= \begin{bmatrix} M_{xx}C^2 + M_{yy}S^2 + 2M_{xy}SC & M_{xy}(C^2 - S^2) + (M_{yy} - M_{xx})SC \\ M_{xy}(C^2 - S^2) + (M_{yy} - M_{xx})SC & M_{xx}S^2 + M_{yy}C^2 - 2M_{xy}SC \end{bmatrix}$$

$$= \mathbf{M}' \tag{2.2.4}$$

Hence the formulae (2.2.1), (2.2.2) can be generated by the matrix similarity transformation (2.2.4).

When engaged in computing the new $\mathbf{M}'$, given $\mathbf{M}$ and θ, the question which (naturally) arises is, are there any values of θ for which $\mathbf{M}_{x'y'} \equiv 0$? Such directions are known as *principal directions* for $\mathbf{M}$, and from (2.2.1, 2) it can be seen that $M_{x'y'} \equiv 0$ when $\theta = \alpha$.

Then

$$\tan 2\alpha = \frac{2M_{xy}}{(M_{xx} - M_{yy})}. \tag{2.2.5}$$

The corresponding $M_{x'x'}$, $M_{y'y'}$ are known as the *principal bending moment values.*

Exercises (1) Note the similarities between $M_{\alpha\beta}$ moment fields and fields of stress and strain in two dimensions.

(2) Suppose in x, y coordinates that at a point in a plate there is pure twist, $M_{xx} = M_{yy} = 0$, $M_{xy} = M_{yx} = m$. Find the principal bending moments, and the planes on which they act.

Now

$$\mathbf{M} = \begin{bmatrix} 0 & m \\ m & 0 \end{bmatrix}.$$

It is required to find a new $\mathbf{M}'$ such that

$$\mathbf{M}' = \begin{bmatrix} \mu_1 & 0 \\ 0 & \mu_2 \end{bmatrix},$$

where $\mu_{1,2}$ are the principal moments. Hint: The appropriate x', y' are inclined at $\pi/4$ to x, y. *Answer:* $\mu_1 = m = -\mu_2$.

2.3 The moment circle

The algebraic manipulations of the previous section can alternatively be carried through to the same result by means of a geometrical construction, the moment circle construction (Fig. 2.5).

The axes for the circle are bending moment (horizontal) and twisting moment (vertical). Consider the planes associated with the x, y axis system, Fig. 2.2. Then the bending moment M_{xx} and twisting moment M_{xy} act on the plane x = const., whereas the bending moment M_{yy} and twisting moment $M_{yx}(=M_{xy})$ act on the plane y = const. Treated as coordinate pairs, two points can now be plotted on the B.M./T.M. axes of the *moment circle*.

One of the M_{xy}s is given a negative sign—suppose it is the one associated with M_{yy}. Then points x,y, as shown on Fig. 2.5, can be plotted. The mid-point of the

$R^2 = M_{xy}^2 + (M_{xx} - M_{yy})^2/4$

$R = (M_{xx} - M_{yy})\operatorname{Sec} 2\alpha/2$

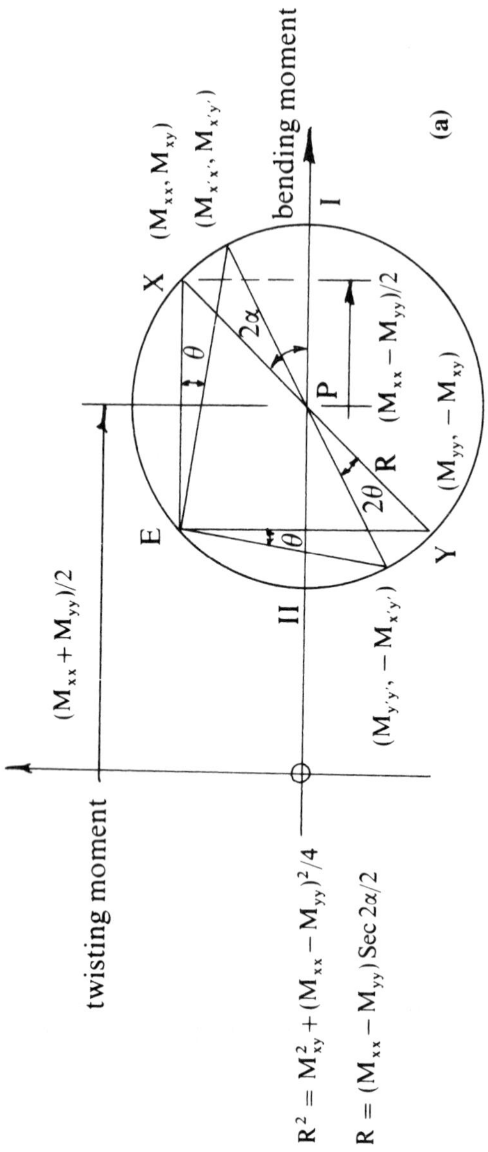

Figure 2.5(a) Moment circle.

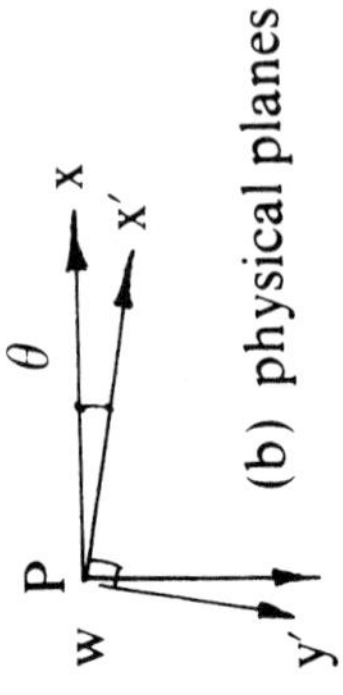

Figure 2.5(b) Physical axes.

line joining x and y is labelled P, and a circle centre P, radius PX(=PY) is drawn. This circle is the moment circle.

By simple geometry on Fig. 2.5 the expression for θ to given principal bending moments can be seen to follow (2.2.5) thus giving the points marked I and II on Fig. 2.5 as the principal bending moments. Note that these bending moments are associated with zero twisting moment, by definition, and that they are algebraically the greatest and least bending moments acting at the point P, selected from amongst all the possible planes drawn through the plate thickness.

Note too that the construction carries data for the *planes* along x and y directions into *points* x and y on the circle which are on the ends of a diameter. Now the planes x and y subtend an angle of $\pi/2$ at P, where the points x and y subtend an angle double this, π, at the circle centre. Likewise the *principal moments*, represented by I, II on the circle act on planes through P which are normal to one another, since these planes must subtend at P, half the angle which I and II subtend at the circle center.

A useful additional feature of the moment circle is the pole point, E, on the perimeter. This is constructed by drawing, through the points x and y respectively, lines which are parallel to the *physical* planes on which the respective moment/twist combinations act. Shown in Fig. 2.5*b* are the physical plane directions—x, y; x′, y′. Now by construction, since the x, y planes are normal, the additional lines now drawn are also normal to one another, and this means that they must intersect *on* the moment circle in order that they may enclose this angle. Then, knowing E, the joins to X′ and Y′ give directions which are parallel to the physical planes on which the M'_{xx}, M'_{xy} etc. act—see also Fig. 2.5*b*.

2.4 Equilibrium equations—rectangular coordinates

Thus far the bending and twisting moments at a point in the plate have been considered, and the question of change of reference axes at the point has been examined. Now an element of the plate is to be considered, and the requirements that this element be in equilibrium will be established. For this a coordinate system must be chosen. Here the simplest system, the rectangular cartesian system, will be adopted. Later another possible choice will be investigated.

Consider the plate element, side lengths dx, dy, as shown in Fig. 2.6. If the plate element is to support any loading, such as a pressure, p(x, y), then not all the internal forces can be constant.

Typical variation of the internal stress resultants is shown in Fig. 2.6. Each of M_{xx}, M_{yy}: M_{xy}, M_{yx}; Q_x, Q_y is shown as varying between the edges of the element, in the manner of the differential calculus. Now apply the equations of statics to this element to determine the equilibrium equations.

Recall first that all the stress resultants are defined per unit length along the middle surface (as seen in plan). Next note that *three* independent equations of

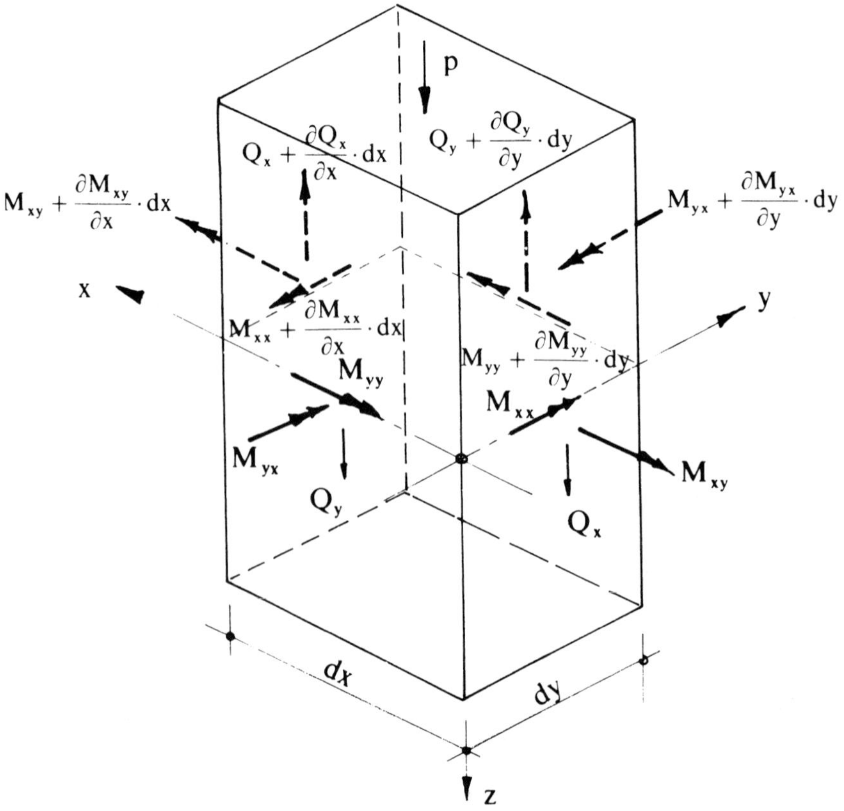

Figure 2.6 Element equilibrium.

statics can be written by choosing *two* distinct axes in the x, y plane about which to take moments, together with a third equation describing the *force* equilibrium in the z direction.

First choose the axis of y about which to take moment equilibrium. Then

$$-M_{xx}.dy+\left(M_{xx}+\frac{\partial M_{xx}}{\partial x}dx\right)dy-M_{yx}dx+\left(M_{yx}+\frac{\partial M_{yx}}{\partial y}dy\right)dx$$
$$-\left(Q_x+\frac{\partial Q_x}{\partial x}dx\right)dy.dx+p.(dx)^2.\frac{dx}{2}=0.$$

Now cancel terms where possible, and note that the terms in $\partial Q_x/\partial x$ and p are of lower order. Note also that there is a common factor of dx . dy, which can be cancelled.

Then the resulting equation of moment equilibrium is

$$\frac{\partial M_{xx}}{\partial x} + \frac{\partial M_{yx}}{\partial y} - Q_x = 0. \tag{2.4.0}$$

By a similar argument, but with respect to the x axis, there is obtained a second moment equilibrium equation

$$\frac{\partial M_{xy}}{\partial x} + \frac{\partial M_{yy}}{\partial y} - Q_y = 0. \tag{2.4.1}$$

The third and final equilibrium equation is for forces in the z direction. Consider, then, the whole element. The forces contributing are the various Q_x, Q_y forces and the applied pressure loading p(x, y).

Thus

$$\left[\left(Q_x + \frac{\partial Q_x}{\partial x}\,dx\right) - Q_x\right]dy + \left[\left(Q_y + \frac{\partial Q_y}{\partial y}\,dy\right) - Q_y\right]dx = p\ dx\,.\,dy.$$

Simplifying, and cancelling by dx . dy,

$$\frac{\partial Q_x}{\partial x} + \frac{\partial Q_y}{\partial y} = p. \tag{2.4.2}$$

These three equations involve five unknowns, M_{xx}, $M_{xy} = M_{yx}$, M_{yy}; Q_x and Q_y. Hence the plate moments are *indeterminate*, in general.

For some purposes it is useful to eliminate the transverse shear forces Q_x, Q_y. When this is done a single equation in the three bending stress resultants is obtained. To obtain this, differentiate (2.4.0) with respect to x, (2.4.1) with respect to y and eliminate Q_x, Q_y between these two and (2.4.2) to give

$$\frac{\partial^2 M_{xx}}{\partial x^2} + 2\frac{\partial^2 M_{xy}}{\partial x \partial y} + \frac{\partial^2 M_{yy}}{\partial y^2} = p. \tag{2.4.3}$$

Note the similarity between this equation and the beam equilibrium equation (1.6.0).

Worked example: Show that the field of moments

$$M_{xx} = -\frac{pL^2}{8}\left(1 - \left(\frac{x}{L}\right)^2\right)$$

$$M_{xy} = \frac{p}{4}xy$$

$$M_{yy} = -\frac{pL^2}{8}\left(1 - \left(\frac{y}{L}\right)^2\right)$$

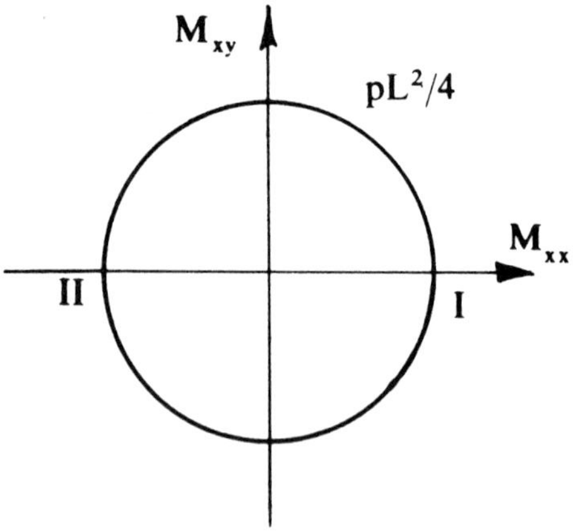

Figure 2.7 Example—moment circle.

is in equilibrium with a pressure loading, p = const., and investigate the principal moments at (O, O), (O, L) and (L, L).

Now

$$\frac{\partial^2 M_{xx}}{\partial x^2} = \frac{p}{4}, \quad \frac{\partial^2 M_{xy}}{\partial x \partial y} = \frac{p}{4}, \quad \frac{\partial^2 M_{yy}}{\partial y^2} = \frac{p}{4}.$$

Hence equation (2.4.3) is satisfied.

At (O, O) $M_{xx} = M_{yy} = -pL^2/8$, $M_{xy} = 0$. Hence M_{xx}, M_{yy} are principal at the origin. At (O, L) $M_{xx} = 0$, $M_{xy} = 0$, $M_{yy} = -pL^2/8$. Hence again M_{xx} and M_{yy} are principal.

At (L, L) $M_{xx} = M_{yy} = 0$ and $M_{xy} = pL^2/4$.

Thus the principal axes are at $\pi/4$ to x, y and the principal moments are $\pm\, pL^2/4$. (See Fig. 2.7).

2.5 Plate bending kinematics—rectangular coordinates

The term kinematics is used to describe the deformations of the plate. When loaded, the plate deflects, carrying the original plane middle surface into some shallow curved shape. This surface will in general have no special features. There will be slopes and curvatures at every point and in every direction through every point on the deflected surfaces. But the deflected shape at the plate will be *shallow* in the sense that everywhere the slopes will be small compared to unity.

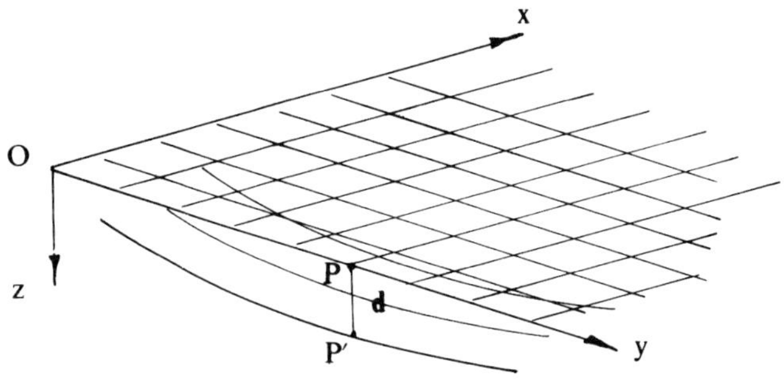

Figure 2.8 Plate deformation—displacement.

As a result, a number of approximations can be introduced to simplify the description of the curved deflected shape of the middle surface. Our purpose here is to decide what geometrical information about this shape will be needed for the construction of physically reasonable plate theories. Recall that it is intended to use a rectangular cartesian coordinate system (Fig. 2.8).

Suppose a typical point (P) on the plate mid surface displaces to P′ as the plate deforms under load. This displacement will be denoted by w(x, y). As a consequence of the shallowness of the deflected surface, the typical point P′ will have no components of displacement in either the x or y directions.

The vector displacement of P to P′ is therefore

$$\mathbf{d} = w\mathbf{k}. \tag{2.5.0}$$

Hence, with respect to an x, y, z coordinate system the deformed middle surface of the plate is described by a radius vector $\mathbf{r}_P$ given by

$$\mathbf{r}_P = x\mathbf{i} + y\mathbf{j} + w\mathbf{k}. \tag{2.5.1}$$

Here x and y are to be regarded as the independent variables—the surface coordinates.

Knowing the form of the radius vector to a typical point **P**, the procedures discussed in the Appendix can now be used to evaluate all the relevant geometrical quantities.

Thus the tangent vectors, $\mathbf{r}_x$, $\mathbf{r}_y$, (see Appendix, section A.2.2) are given by

$$\mathbf{r}_x \equiv \frac{\partial \mathbf{r}}{\partial x} = \mathbf{i} + \frac{\partial w}{\partial x} . \mathbf{k},$$

$$\mathbf{r}_y \equiv \frac{\partial \mathbf{r}}{\partial y} = \mathbf{j} + \frac{\partial w}{\partial y} . \mathbf{k},$$

and the coefficients of the first fundamental form, E, F, G, (A.2.4) are given by

$$E \equiv \mathbf{r}_x \cdot \mathbf{r}_x = 1 + \left(\frac{\partial w}{\partial x}\right)^2,$$
$$F \equiv \mathbf{r}_x \cdot \mathbf{r}_y = \left(\frac{\partial w}{\partial x} \cdot \frac{\partial w}{\partial y}\right), \tag{2.5.2}$$
$$G \equiv \mathbf{r}_y \cdot \mathbf{r}_y = 1 + \left(\frac{\partial w}{\partial y}\right)^2.$$

Recall now the notion of shallowness; this implies that

$$\left(\frac{\partial w}{\partial x}\right), \left(\frac{\partial w}{\partial y}\right) \ll 1 \quad \text{then}$$

$$E \doteqdot 1, \quad F \doteqdot 0, \quad G \doteqdot 1. \tag{2.5.3}$$

These approximate expressions indicate that the original coordinate lines on the deformed plate middle surface are still sensibly *orthogonal.*

Proceed to consider the unit surface normal (A.3) and the surface curvatures. Now **n**, the unit normal, is defined by (A.3.0) as

$$\mathbf{n} \equiv \frac{\mathbf{r}_x \times \mathbf{r}_y}{|\mathbf{r}_x \times \mathbf{r}_y|}.$$

In this case

$$\mathbf{r}_x \times \mathbf{r}_y = -\frac{\partial w}{\partial x}\mathbf{i} - \frac{\partial w}{\partial y}\mathbf{j} + 1.\mathbf{k},$$

and

$$|\mathbf{r}_x \times \mathbf{r}_y| = \left[1 + \left(\frac{\partial w}{\partial x}\right)^2 + \left(\frac{\partial w}{\partial y}\right)^2\right]^{1/2} \doteqdot 1.$$

Hence

$$\mathbf{n} = -\frac{\partial w}{\partial x}\mathbf{i} - \frac{\partial w}{\partial y}\mathbf{j} + \mathbf{k}. \tag{2.5.4}$$

Next the coefficients of the second fundamental form (A.4.1) are given by

$$L \equiv \mathbf{n} \,.\, \mathbf{r}_{xx} = \left(-\frac{\partial w}{\partial x}\mathbf{i} - \frac{\partial w}{\partial y}\mathbf{j} + \mathbf{k}\right) . \left(\frac{\partial^2 w}{\partial x^2}\mathbf{k}\right),$$

$$= \frac{\partial^2 w}{\partial x^2}$$

$$M \equiv \mathbf{n} \,.\, \mathbf{r}_{xy} = \left(-\frac{\partial w}{\partial x}\mathbf{i} - \frac{\partial w}{\partial y}\mathbf{j} + \mathbf{k}\right) . \left(\frac{\partial^2 w}{\partial x \partial y}\mathbf{k}\right), \qquad (2.5.5)$$

$$= \frac{\partial^2 w}{\partial x \partial y}$$

and

$$N \equiv \mathbf{n} \,.\, \mathbf{r}_{yy} = \left(-\frac{\partial w}{\partial x}\mathbf{i} - \frac{\partial w}{\partial y}\mathbf{j} + \mathbf{k}\right) . \left(\frac{\partial^2 w}{\partial y^2}\mathbf{k}\right),$$

$$= \frac{\partial^2 w}{\partial y^2}.$$

Hence finally, recalling that the surface coordinate lines namely the x, y coordinates projected on to the deflected middle surface, are sensibly orthogonal, therefore the curvatures and twists on the deformed middle surface are given by

$$\kappa_{xx} \equiv L/E = \frac{\partial^2 w}{\partial x^2}, \qquad (A.9.0)$$

$$\kappa_{xy} \equiv k_{yx} \equiv M/\sqrt{EG} = \frac{\partial^2 w}{\partial x \partial y}, \qquad (A.9.4) \qquad (2.5.6)$$

$$\kappa_{yy} \equiv N/G = \frac{\partial^2 w}{\partial y^2}. \qquad (A.9.1)$$

These expressions are the basic kinematic relations required for the construction of a suitable plate theory when expressed in cartesian coordinates.

Worked example: Consider the deflected shape

$$w(x, y) = A \sin\frac{\pi x}{L} \sin\frac{\pi y}{L},$$

to be defined over the square $0 \leqslant x \leqslant L$, $0 \leqslant y \leqslant L$. Then $w = 0$ all round the boundary. At a typical point x, y

$$\kappa_{xx} = -\left(\frac{\pi}{L}\right)^2 w(x, y) = \kappa_{yy} \quad \text{and} \quad \kappa_{xy} = A\left(\frac{\pi}{L}\right)^2 \cos\frac{\pi x}{L} \cos\frac{\pi y}{L}.$$

Consider the direction x = y, that is, a diagonal of the square. At the centre $\kappa_{xy} = 0$ since $x = y = L/2$, then $\kappa_{xx} = \kappa_{yy}$ are *principal curvatures*, and are at maximum value. At $x = y = L/4$, then $\kappa_{xx} = \kappa_{yy} = -A/2(\pi/L)^2$ and $\kappa_{xy} = +A/2(\pi/L)^2$. The curvature circle has a vertical diameter through the centre, with the x, y directions at the ends. Hence the x, y directions are always at $\pi/4$ to the principal curvature directions, except when $\kappa_{xy} = 0$, when the x, y directions are principal—and so too are all other directions.

Exercise: If $w = Ax(a-x)y(b-y)$ as $0 \leqslant x \leqslant a$, $0 \leqslant y \leqslant b$, derive the curvature matrix $\boldsymbol{\kappa}$ (A.10.0) and explore the properties of the principal directions.

Answer: $\kappa_{xx} = -2Ay(b-y)$, $\kappa_{xy} = A\,(a-2x)(b-2y)$.

2.6 Equilibrium equations—polar coordinates—radial symmetry

Earlier, equilibrium of a plate element referred to rectangular coordinates was presented. Here polar coordinates will be considered. However only a special case will be dealt with, the case of axial symmetry. This symmetry corresponds to independence for all the dependent quantities from change of the angular coordinate, θ—termed *axial symmetry*. This feature has been incorporated in the stress resultant array on Fig. 2.9.

Since only variation with respect to r is now possible, so ordinary rather than partial differentials are indicated. Also, this form of symmetry ensures that $Q_\theta = 0$. It is left as an exercise for the reader to construct and demonstrate the correctness of Fig. 2.9.

Because of the symmetry, only two independent equilibrium equations exist—one a moment equation and the other a force equation for resolution in the axial direction.

Consider moments of the forces shown in Fig. 2.9 about a tangential axis. Then

$$\left[M_{rr} + \frac{dM_{rr}}{dr}dr\right][r + dr]d\theta - M_{rr}\,.\,rd\theta - M_{\theta\theta}.2\frac{d\theta}{2}dr$$

$$-\left[Q_r + \frac{dQ_r}{dr}dr\right][r + dr]d\theta\,.\,dr = 0.$$

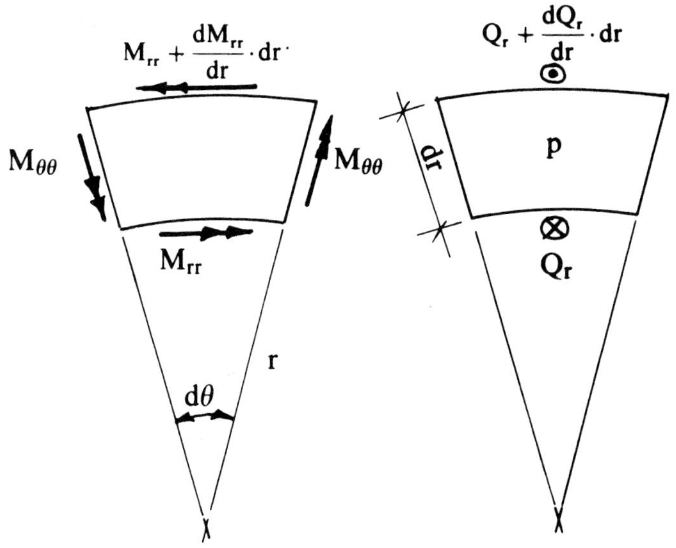

Figure 2.9 Circular plate—element forces.

Simplify and neglect the second order term in dQ_r/dr, then divide through by $dr \, . \, d\theta$.

Hence there is obtained

$$r\frac{dM_{rr}}{dr} + M_{rr} - M_{\theta\theta} - Q_r \, . \, r = 0. \tag{2.6.0}$$

By considering force equilibrium in the z-direction, obtain

$$\left[Q_r + \frac{dQ_r}{dr} \cdot dr\right][r + dr]d\theta - Q_r \cdot r \cdot d\theta - p \cdot r \cdot d\theta \cdot dr = 0,$$

or

$$r\frac{dQ_r}{dr} + Q_r = p \, . \, r \, . \tag{2.6.1}$$

These two equations can be rewritten as

$$\begin{aligned} (rM_{rr})_{,r} - M_{\theta\theta} &= r \, . \, Q_r, \\ (rQ_r)_{,r} &= p \, . \, r \, . \end{aligned} \tag{2.6.2}$$

These are two equations in three unknowns, M_{rr}, $M_{\theta\theta}$,Q_r and hence are indeterminate.

Exercise: Show that $M_{rr} = p\,(a^2 + r^2)/6$, $M_{\theta\theta} = pa^2/6$, is an equilibrium system with p = const., a = const., and find Q_r. *Answer:* $Q_r = pr/2$.

2.7 Plate bending kinematics—polar coordinates—radial symmetry

Useful theories for plate behaviour can only be constructed once both statics and kinematics are linked, because of the statical indeterminacy. In this section, the basic aim is to evaluate the deflected shape *curvatures* from a knowledge of the deflected shape itself, and in the following chapters complete theories will be discussed.

As for the rectangular coordinate description, and resulting from the shallow character of the deflected surface, the typical radius vector for a point on the plate deflected surface can now be written as

$$\mathbf{r} = r\,.\,\mathbf{I} + w(r)\,.\,\mathbf{K} \qquad (2.7.0)$$

where the position vector **r** is seen to depend only on r and the transverse deflection, w. The unit vectors **I** and **K** are directed radially and transversely—see Fig. 2.10.

The deflected middle surface in this case is a surface of revolution, because radial symmetry is assumed. Hence the results of Appendix A.13 are relevant. However, since the surface here is a *shallow* shape, the curvature expressions can conveniently be evaluated from first principles. Now the surface slope in the radial direction at P is given by $w,_r$. From symmetry, the circumferential slope is zero.

It is reasonable to suppose that the section (Fig. 2.10*a*) cuts the deflected surface along a principal curvature direction. This is confirmed by the analysis of section A.13. The curvature in this plane, κ_1, is associated with a centre of curvature O_1, and a radius O_1P (see inset, Fig. 2.10*a*). The second principal curvature is then associated with the section of the surface by a plane containing the normal at P, and the normal to the plane of the Section A–A. As shown in Section A.13 this curvature is associated with a centre of curvature O_2 which is given by the point where the normal intersects the axis of revolution.

Take the surface coordinates to be r and θ. By this is meant that the point P is located in plan by the choice at r and θ, then P′, the point required on the deflected surface is vertically below P, a distance w(r), the amount by which the point originally at P deflects to P′.

Now, the operation $(..),_r$ does not affect any of the orthogonal unit vectors **I**, **J**, **K**; however, $(..),_\theta$ does.

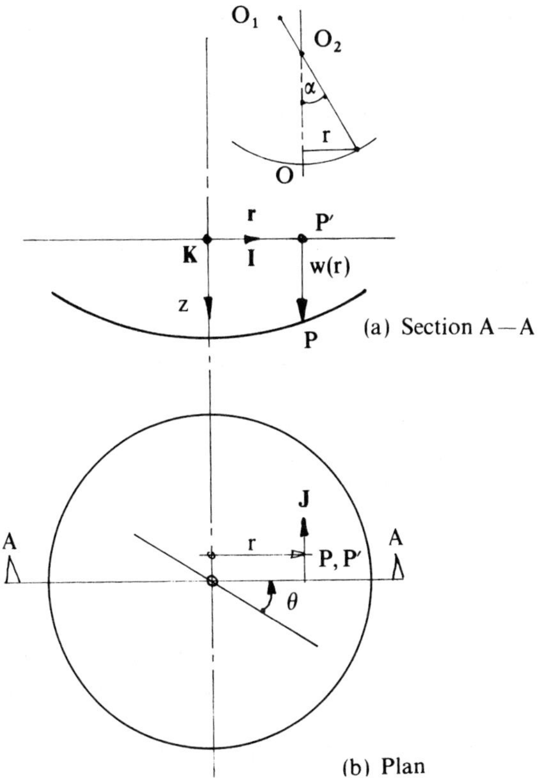

Figure 2.10 Circular plate–deformation geometry.

Thus

$$\mathbf{I},_\theta = \mathbf{J}, \quad \mathbf{J},_\theta = -\mathbf{I}, \qquad \mathbf{K},_\theta = 0. \tag{2.7.1}$$

Hence,

$$\mathbf{r},_r = \mathbf{I} + w,_r\mathbf{K},$$
$$\mathbf{r},_\theta = r\mathbf{J} + O.$$

The coefficients of the first fundamental form (A.2.4) are then

$$E = \mathbf{r},_r \,.\, \mathbf{r},_r = 1 + w,_r^2 \doteqdot 1$$
$$F = \mathbf{r},_r \,.\, r,_\theta = 0,$$
$$G = r^2.$$

Now

$$\mathbf{n} = \frac{\mathbf{r},_r \times \mathbf{r},_\theta}{\left|\mathbf{r},_r \times \mathbf{r},_\theta\right|} = \frac{-r \,.\, w,_r \,.\, \mathbf{I} + r \,.\, \mathbf{K}}{\left(r^2 + r^2 w,_r^2\right)^{1/2}}$$

but $w,_r \ll 1$, hence

$$\mathbf{n} = -w,_r \,.\, \mathbf{I} + \mathbf{K}.$$

Then the second fundamental form coefficients (A.4.0, A.4.1) are

$$L \equiv \mathbf{n} \,.\, r,_{rr} = \mathbf{n} \,.\, (w,_{rr})\mathbf{K},$$

$$= w,_{rr};$$

$$M \equiv \mathbf{n} \,.\, \mathbf{r},_{r\theta} = n \,.\, J = 0;$$

$$N \equiv \mathbf{n} \,.\, \mathbf{r},_{\theta\theta} = n. -r\mathbf{I} = rw,_r.$$

The r, θ directions are therefore principal, because for this choice of coordinates $F = M = 0$.

Hence the principal curvatures (A.10.1) are given by

$$\kappa_1 = L/E = w,_{rr},$$

$$\kappa_2 = N/G = \frac{w,_r}{r}. \qquad (2.7.2)$$

The curvature, κ_1, is that associated with the radial section, Fig. 2.10*a*, and the curvature centre O_1, while the curvature κ_2 is the curvature of a section transverse to the previous one and is associated with the curvature centre O_2.

These curvature values can be confirmed by direct calculation using the properties of O_1 and O_2.

Thus from Fig. 2.10 (inset), since the curve OP is shallow, the curvature of OP at P is given by $w,_{rr}$ as for a beam. Also from the figure, $O_2P = r/\sin\alpha$ and α is small, being such that $\tan\alpha = w,_r \ll 1$. Hence $\sin\alpha = \tan\alpha = w,_r$, and $O_2P = r/w,_r$, or the curvature

$$\kappa_2 = \frac{1}{O_2P} = w,_r/r.$$

Exercise: Suppose $w = C(a^2 - r^2)^2$, a = const., show that the principal curvatures are given by

$$\kappa_{rr} = 4C \,.\, (3r^2 - a^2), \qquad \kappa_{\theta\theta} = 4C \,.\, (r^2 - a^2).$$

2.8 Conclusions

The ingredients for suitable plate theories have been identified in this chapter. These are equilibrium equations and deformation (kinematic) relations. The two simplest and most useful coordinate systems have been considered, namely rectangular cartesian and radially symmetric polar. In the next three chapters these ingredients will be blended to provide useful plate theories.

3 Elastic plates

3.0 Introduction

The earlier chapters have dealt with statics and kinematics of plate bending. That is, they have dealt with the conditions which the internal forces and moments must satisfy if they are to be in equilibrium with the loads; also, the methods have been developed for describing the deformation of the plate in its bent state. Before actual problems of plate bending can be formulated and solved, these two aspects have to be brought together, and in this chapter this is done for the case of the elastic plate. Roughly speaking, plates respond elastically in the working load condition, and plastically at the ultimate or collapse state. This chapter will be dealing with plates at working load, and the next will deal with plastic plates, namely plates in the final stages of loading before they collapse.

All the plates considered in this chapter will be assumed to be of constant thickness and isotropic, i.e. material properties not varying with direction in the plate.

3.1 Elastic theory of plate bending — moment/curvature relations

For simplicity, the coordinate system to be used will be assumed to be rectangular cartesian.

An elastic plate is characterized by a linear relation between load and deflection. However, the material response is not so much related to *deflection* as to *curvature change*. Recall the elastic beam, where the material response is a linear relation between bending moment and curvature change. The same is true for the plate, though now there is a two-dimensional aspect to the problem. The plate under load will develop internal force systems of bending moments (including twisting moments) and transverse shear forces to resist the loading and the appropriate equilibrium equations must also be satisfied.

In addition, the plate will deform, and the originally plane plate will deflect and develop curvatures and twists. For an elastic plate in bending, the bending moments relate linearly to the curvatures. The deformation of the plate has been characterized by the deformation of the middle surface. In the plate, the material on one side of the middle surface will be compressed, that on the other will be extended.

The plates with which we are concerned will be assumed to deform according to the *Kirchhoff hypothesis*. This is the name associated with a mechanical model of behaviour, in which the plate can be thought of as composed of a series of laminae, as in plywood. One of these, the middle one, is the middle surface which can deform by bending but is not extended or compressed in its own plane. If the other plys are thought of as secured by nails through the thickness, then the Kirchhoff hypothesis states that these nails are not extended or sheared by the deformation. Expressed in other words, the plane sections taken normal to the middle surface before deformation remain plane, and normal to the middle surface and unextended after deformation (Fig. 3.0).

Consider then the analysis of this model of behaviour. The requirement is to evaluate the strain in the x-direction at transverse distance z from the un-strained middle surface. If the transverse displacement is denoted by w(x, y) then the slope in the x-direction will be $w,_x$ and

$$\begin{aligned}\varepsilon_{xx} &= \frac{z(w,_x + w,_{xx}dx) - z(w,_x)}{dx} \\ &= zw,_{xx}, \text{ here compression.}\end{aligned} \tag{3.1.0}$$

In a similar manner

$$\varepsilon_{yy} = zw,_{yy} \quad \text{and} \quad \varepsilon_{xy} = zw,_{xy}.$$

But from the earlier discussion of middle surface curvatures it is seen that

$$\varepsilon_{xx} = z\kappa_{xx}, \varepsilon_{xy} = z\kappa_{xy}, \varepsilon_{yy} = z\kappa_{yy}. \tag{3.1.1}$$

Now the elastic response in this typical ply of the plate is

$$E.\varepsilon_{xx} = Ez\kappa_{xx} = \sigma_{xx} - \nu\sigma_{yy} - \nu\sigma_{zz} \tag{3.1.2}$$

and to a good approximation $\sigma_{zz} = 0$. (Here E is Young's modulus).

Recalling the definitions for stress resultants suggests that (3.1.2) should be multiplied by z and then an integration through the plate thickness performed.

Thus

$$\int_{-t/2}^{t/2} E.z^2.\kappa_{xx}dz = \int_{-t/2}^{t/2} \sigma_{xx}.z\,dz - \nu \int_{-t/2}^{t/2} \sigma_{yy}.z\,dz.$$

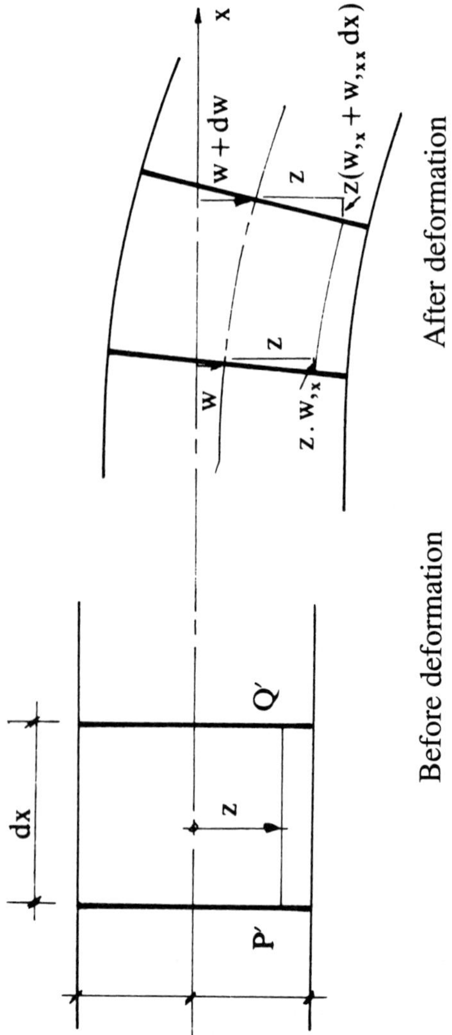

Figure 3.0 Deformation geometry.

Then noting that κ_{xx} is not a function of z,

$$\frac{Et^3}{12}\kappa_{xx} = M_{xx} - \nu M_{yy}.$$

In a similar manner

$$\frac{Et^3}{12}.\kappa_{yy} = M_{yy} - \nu M_{xx}.$$

Now solve for M_{xx} and M_{yy}; then on multiplication by ν and adding, M_{yy} can be eliminated and M_{xx} found.

Hence

$$M_{xx} = D(\kappa_{xx} + \nu\kappa_{yy}), D \equiv \frac{Et^3}{12(1-\nu^2)},$$

and

$$M_{yy} = D(\kappa_{yy} + \nu\kappa_{xx}).$$

In a similar manner, for each ply being sheared

$$2G.\varepsilon_{xy} = \sigma_{xy}$$

where G is the shear modulus.

By a similar process there is obtained

$$M_{xy} = \frac{2Gt^3}{12}\kappa_{xy} = D(1-\nu).\kappa_{xy}.$$

Hence collecting the elastic material relations together gives

$$\begin{bmatrix} M_{xx} \\ M_{yy} \\ M_{xy} \end{bmatrix} = D\begin{bmatrix} 1 & \nu & 0 \\ \nu & 1 & 0 \\ 0 & 0 & 1-\nu \end{bmatrix}\begin{bmatrix} \kappa_{xx} \\ \kappa_{yy} \\ \kappa_{xy} \end{bmatrix}. \tag{3.1.3}$$

Notes: The relations (3.1.3) summarize the elastic material response. They relate bending and twisting moments to *changes* in curvature and twist for the deformed middle surface.

Exercise: Find the bending and twisting moments associated with the deflected shapes defined in the two examples in 2.5. *Answer:*

$$M_{xx} = -2AD[y(b-y) + \nu x(a-x)], M_{xy} = AD(1-\nu)(a-2x)(b-2y).$$

3.2 Elastic theory of plate bending—the governing equation (cartesian coordinates)

The ingredients in the elastic theory of plate bending are the moments/unit length M_{xx}, M_{yy}, M_{xy} (3 quantities), the transverse shear forces/unit length Q_x, Q_y (2 quantities) and the middle surface changes of curvature and twist κ_{xx}, κ_{yy}, κ_{xy} (3 quantities). These eight quantities are dependent, and are generally unknown, and have to be found by solving the particular problem. The loading p(x, y) will be assumed specified, and all quantities are functions of x and y.

Eight equations are required to connect the eight dependent quantities, but there is a further very important dependent quantity, w(x, y), the transverse displacement, to be considered, so there are nine dependent quantities in all, and thus nine equations are required: these are three equilibrium equations, three moment-curvature relations and three relations expressing curvatures in terms of the displacement, w.

Now in (2.4.3) it has been shown that plate bending equilibrium can be expressed by the single equation

$$M_{xx,xx} + 2M_{xy,xy} + M_{yy,yy} = p. \qquad (3.2.0)$$

If the relations (3.1.3) are substituted for the stress resultants then (3.2.0) becomes

$$D[(\kappa_{xx} + \nu\kappa_{yy}),_{xx} + 2(1-\nu)\kappa_{xy,xy} + (\kappa_{yy} + \nu\kappa_{xx}),_{yy}] = p. \qquad (3.2.1)$$

Recalling that $\kappa_{xx} = w,_{xx}$, $\kappa_{xy} = w,_{xy}$, $\kappa_{yy} = w,_{yy}$, then (3.2.1) can be written entirely in terms of w as

$$D[w,_{4x} + \nu w,_{2x2y} + 2(1-\nu)w,_{2x2y} + w,_{4y} + \nu w,_{2x2y}] = p.$$

The ν terms cancel, whence

$$w,_{4x} + 2w,_{2x2y} + w,_{4y} = p/D. \qquad (3.2.2)$$

This equation in the single dependent variable w is usually written in shorthand as

$$\nabla^4 w = p/D. \qquad (3.2.3)$$

This is the celebrated *Biharmonic equation* and is *the governing equation* for elastic plate bending analyses.

Notes: (1) When comparing plate theory with beam theory it can be seen that there is a formal resemblance between

$$EIw^{IV} = p \text{ (see (1.6.1))},$$

and

$$D\nabla^4 w = p \text{ (see (3.2.3))}.$$

Indeed, this resemblance is quite useful to recall from time to time, and is an aid to anticipating plate behaviour by comparing with beam equivalents. The main differences stem from the twisting action which can be developed by plates but not by beams.

(2) The biharmonic is a *linear* equation and hence superposition of solutions is possible.

3.3 Circular plates—radial symmetry

The governing equation (3.2.3), the biharmonic equation, has been derived using the rectangular cartesian coordinate system. There will be occasions when other coordinate systems are needed, particularly the polar system. Here the plate equations for the simplest polar system, with radial symmetry, will be described.

In the previous chapter the equations of equilibrium were derived, so too were expressions for the curvatures (2.6, 2.7). The curvatures, when allied with the Kirchhoff hypothesis of plane sections remaining plane give the link between forces on the one hand and strains and deformations on the other.

Thus the direct strain in the r-direction at a transverse distance z from the middle plane is given by

$$\varepsilon_{rr} = z\,.\,\kappa_{rr} = zw,_{rr}\,. \tag{3.3.0}$$

Again, in the transverse direction

$$\varepsilon_{\theta\theta} = z\,.\,\kappa_{\theta\theta} = z\frac{w,_r}{r}\,.$$

There is no shear strain because of the assumed radial symmetry. These strains then are those experienced by a layer of the plate parallel to the middle surface. As earlier in this chapter, using 3.3.0 in the elastic relations then multiplying by z and integrating through the thickness, the required stress resultants are recovered and suitable moment-deformation relations are derived.

Instead of this fundamental approach here it is observed that when this procedure is carried out in two dimensions the following type of relations result. If the directions of the coordinates are denoted by $(..)_1$ and $(..)_2$, and the coordinates are orthogonal, then

$$M_{11} = D(\kappa_{11} + \nu\kappa_{22}),$$

and

$$M_{22} = D(\kappa_{22} + \nu\kappa_{11}), \tag{3.3.1}$$

where

$$D = \frac{Et^3}{12(1-\nu^2)},$$

E = Young's Modulus, ν = Poisson's Ratio and t is the plate thickness.

Here then

$$M_{rr} = D\left(w,_{rr} + \nu\frac{w,_r}{r}\right), \qquad M_{\theta\theta} = D\left(\frac{1}{r}w,_r + \nu w,_{rr}\right). \tag{3.3.2}$$

Now take stock of what dependent variables are present, and how many equations relate. The dependent quantities are M_{rr}, $M_{\theta\theta}$ Q_r; κ_{rr}, $\kappa_{\theta\theta}$ and w; six in all. The available equations are two equilibrium (2.6.2), two stress-strain (3.3.2) and two curvature displacement (2.7.2). Hence there are six equations and six unknowns. The preferred unknown, as for rectangular coordinates, is w, the transverse displacement. The equations are

$$\begin{aligned}(rM_{rr}),_r - M_{\theta\theta} &= r\,.\,Q_r,\\ (rQ_r),_r &= p\,.\,r,\\ M_{rr} &= D(\kappa_{rr} + \nu\kappa_{\theta\theta}),\\ M_{\theta\theta} &= D(\kappa_{\theta\theta} + \nu\kappa_{rr}),\\ \kappa_{rr} &= w,_{rr},\\ \kappa_{\theta\theta} &= \frac{1}{r}w,_r.\end{aligned} \tag{3.3.3}$$

The equations are structured in such a way that the three forces and moments are described by *only two* equilibrium equations—that is, the circular plate is a statically indeterminate problem.

The simplest procedure for solution is to derive the analogous equation to (3.2.3); indeed this equation will be (3.2.3) expressed in polar coordinates. To achieve this, first eliminate Q_r between the equilibrium equations, thus

$$(rM_{rr}),_{rr} - M_{\theta\theta},_r = p\,.\,r. \tag{3.3.4}$$

Then eliminate M_{rr}, $M_{\theta\theta}$ by use of the moment-curvature relations to obtain

$$(r\kappa_{rr} + \nu\kappa_{\theta\theta}),_{rr} - (\kappa_{\theta\theta} + \nu\,.\,\kappa_{rr}),_r = \frac{pr}{D}. \tag{3.3.5}$$

Finally, substitute the values of the curvatures in terms of the displacement derivatives, when

$$(rw,_{rr} + \nu w,_r),_{rr} - \left(\frac{1}{r}w,_r + \nu w,_{rr}\right),_r = \frac{pr}{D}.$$

Simplifying by expanding terms and cancelling, when

$$(rw,_{rrr} + w,_{rr})_r + \nu(w,_{rrr}) - \frac{1}{r}w,_{rr} + \frac{1}{r^2}w,_r - \nu w,_{rrr} = \frac{pr}{D},$$

or

$$w,_{4r} + \frac{2}{r}w,_{3r} - \frac{1}{r^2}w,_{2r} + \frac{1}{r^3}w,_r = \frac{p}{D}. \qquad (3.3.6)$$

This equation can be written in another form which is more easily used to construct a solution. Thus (3.3.6) can be grouped and then written as

$$\frac{1}{r}\left[r\left(\frac{1}{r}(rw,_r),_r\right),_r\right],_r = \frac{p}{D}. \qquad (3.3.7)$$

This second form then allows an integration in quadratures, that is, repeated direct integration.

Note: As remarked earlier, (3.3.6 and 7) are in reality (3.2.3) expressed in polar coordinates but with all variation with respect to θ suppressed.

3.4 Some simple solutions for circular plates

Following up the suggestion in the previous section, (3.3.7) will be integrated by repeated integration. Thus multiply through by r and then integrate with respect to r, when

$$r\left(\frac{1}{r}(rw,_r),_r\right),_r = \frac{pr^2}{2D} + \bar{A}.$$

Now divide by r and carry out a further integration, when

$$\frac{1}{r}(rw,_r),_r = \frac{pr^2}{4D} + \bar{A}\ln r + \bar{B}.$$

Next multiply by r and integrate a third time, then

$$(rw,_r) = \frac{pr^4}{16D} + \bar{A}\left[\ln r \,.\, \frac{r^2}{2} - \int \frac{r^2}{r} \,.\, \frac{1}{r}\,dr\right] + \frac{\bar{B}r^2}{2} + C.$$

Finally

$$w = \frac{pr^4}{64D} + \frac{\bar{A}r^2}{8}[2\,.\ln r - 1] + \frac{\bar{B}r^2}{2} + \bar{C}\,.\ln r + \bar{D}.$$

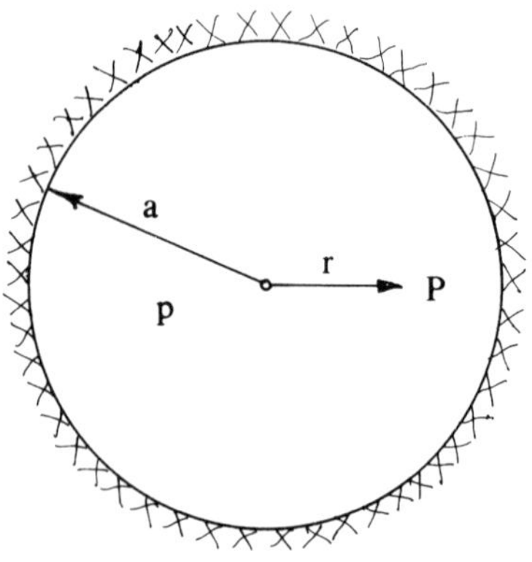

Figure 3.1 Circular clamped plate—uniform loading.

Defining new arbitrary constants, which is a convenience rather than a necessity, the general solution of (3.3.6) and (3.3.7) can be seen to be

$$w = \frac{pr^4}{64D} + A \,.\, \ln r + Br^2 \,.\, \ln r + Cr^2 + E. \tag{3.4.0}$$

This solution consists of a complementary function, namely the portion containing the four arbitrary constants A, B, C and E together with the *particular integral* for p, which has been evaluated here on the assumption that p is a constant. If p is some function of r then suitable particular integrals are readily calculated, exactly in the manner of the present one, if required. Also, as for beams, so here, loadings such as point loads cannot be directly represented in the solution, since they represent discontinuous happenings in the member, and (3.4.0), being the regular solution of an ordinary differential equation (3.3.7), is written explicitly for *full continuity*. Discontinuities can be handled but this is a specialized point which will not be considered here.

Consider an example of a uniform, clamped-edge circular plate, radius a, with a uniform distribution of applied loading, p (Fig. 3.1). Find an expression for the deflected shape and evaluate the bending moments. The starting point is the general expression for the transverse displacement, (3.4.0).

Hence

$$w = \frac{pr^4}{64D} + A \,.\, \ln r + Br^2 \,.\, \ln r + Cr^2 + E.$$

As with beam problems, so here the search for conditions to fix the values for the (four) arbitrary constants must be sought from two conditions at each end of the range of integration. The range of integration here is $0 \leqslant r \leqslant a$.

At r = 0 the requirement is that w be finite and $w,_r = 0$. These conditions can be met only if A = B = 0, since ln r is singular at r = 0. These then are the two conditions at r = 0.

At r = a, the conditions are that w(a) = 0, $w,_r(a) = 0$. From the general expression for w, then

$$C.a^2 + E = -\frac{pa^4}{64D},$$

and

$$2C.a = -\frac{pa^3}{16D}.$$

Hence

$$C = -\frac{pa^2}{32D}, E = \frac{pa^4}{64D}.$$

Thus all four constants have been found and

$$w(r) = \frac{pr^4}{64D} - \frac{pa^2r^2}{32D} + \frac{pa^4}{64D}$$

$$= \frac{p}{64D}(a^2 - r^2)^2. \quad (3.4.1)$$

This then is the shape taken up by the plate for a uniform loading.

To evaluate the bending moments, it is convenient to first find the curvatures from (3.4.1).

Now

$$w,_{rr} (= \kappa_1) = \frac{p}{16D}[-a^2 + 3r^2],$$

and

$$\frac{1}{r} w,_r (= \kappa_2) = \frac{p}{16D}[-a^2 + r^2].$$

Hence

$$M_{rr} = D\left(w,_{rr} + \nu \frac{1}{r} w,_r \right) = +\frac{pr^2}{16}(3+\nu) - \frac{pa^2}{16}(1+\nu),$$

and

$$M_{\theta\theta} = D\left(\frac{1}{r} w,_r + \nu w,_{rr} \right) = +\frac{pr^2}{16}(1+3\nu) - \frac{pa^2}{16}(1+\nu).$$

For small r both M_{rr} and $M_{\theta\theta}$ are negative, and hence are *sagging* bending moments. Near the edge both moments are positive and hence are *hogging* bending moments. At $r=\sqrt{(1+\nu)/(3+\nu)}\,.\,a$, M_{rr} changes sign, while at $r=\sqrt{(1+\nu)/(1+3\nu)}\,.\,a$, $M_{\theta\theta}$ changes sign.

Notes: (1) *Sagging* bending moment causes tension on the underside of the slab, *hogging* bending moment produces top tension.

(2) The form of the governing equation (3.3.6) can be integrated without knowledge of the form (3.3.7), but it is essentially a trial process. For example, suppose the solution is sought in the form $w = Ar^n$, then the terms in r^2 and a constant can be found, but not the other two involving ln r.

(3) If the radius r = 0 is contained in the plate, that is if there is no hole at the centre, then for finiteness at r = 0, A = B = 0 must follow. No further conditions at r = 0 can then be applied—for example a zero slope requirement from symmetry. Such conditions are implied by A = B = 0, and the remaining conditions at the outer radius.

As a second example consider the clamped circular plate but loaded with a central point load W rather than a uniform distribution of loading (Fig. 3.2). Find the deflected shape, and distribution of bending moments.

The procedure is to regard the point load as being distributed over a small circular area at the centre of the plate. Suppose the radius of this area to be δ. The distribution of transverse shear force Q_r then can be found from that which equilibrates this load.

Thus,

$$2\pi\delta Q_r = W.$$

Actually, δ can be any radius r and it follows that $rQ_r = W/2\pi$. Since $p \equiv 0$ in this case,

$$w = A\,.\,\ln r + Br^2\,.\,\ln r + Cr^2 + E.$$

Then

$$w,_r = \frac{A}{r} + B(2r\,.\,\ln r + r) + 2Cr.$$

and

$$w,_{rr} = -\frac{A}{r^2} + B(2r\,.\,\ln r + 3) + 2C.$$

Now M_{rr} and $M_{\theta\theta}$ can be evaluated, and so

$$r\,.\,Q_r = (rM_{rr}),_r - M_{\theta\theta} = 4D\,.\,B.$$

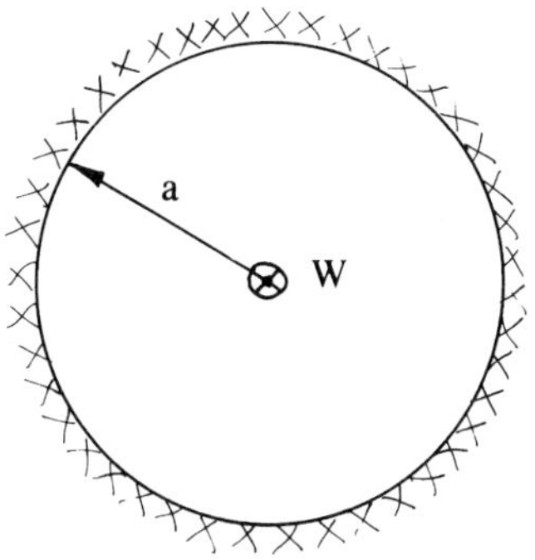

Figure 3.2 Example, circular clamped plate—point loading.

Hence

$$B = \frac{rQ_r}{4D} = \frac{W}{8\pi D}.$$

The remaining conditions are slope = 0 at the origin,

$$(A = 0),\ w(a) = 0 \quad \text{and} \quad w,_r(a) = 0.$$

Hence

$$E = \frac{Wa^2}{16\pi D}, \quad C = -\frac{W}{16\pi D}(2\ln a + 1).$$

Finally,

$$w(r) = \frac{W}{8\pi D}\left[r^2.\ln\frac{r}{a} - \frac{1}{2}(r)^2 + \frac{a^2}{2}\right].$$

Hence the central deflection is

$$w(0) = \frac{Wa^2}{16\pi D}.$$

Exercise: Consider the plate of 3.4, but now with the edge simply supported rather than clamped. The conditions at the (outer) edge will then be that the displacement and radial bending moment are zero. For simplicity assume Poisson's ratio (ν) to be zero, and show that

$$w(r)=\frac{p}{64D}(a^2-r^2)(5a^2-r^2),$$

$$M_{rr}(r)=-\frac{3p}{16}(a^2-r^2),$$

$$M_{\theta\theta}(r)=-\frac{p}{16}(3a^2-r^2).$$

Can you devise a check on these results by use of the clamped edge results?

3.5 Simple solutions for problems in rectangular coordinates

The quest now is for solutions of the biharmonic equation, (3.2.3). The simplest useful problems relate to rectangular plates.

A basic method of solution and really the only feasible formal method for solution of engineering plate problems, is the method of *separation of variables*. There are other methods, considered later, but these are essentially numerically based, or approximate methods.

The method of separation of variables relates mostly to plates of rectangular shape, and begins by assuming that w(x, y), the dependent quantity being sought, can be written as a *product* of two separate functions of X(x) and of Y(y).

Thus the solution process begins by letting

$$w(x, y) = X(x) \,.\, Y(y). \tag{3.5.0}$$

Usually this is not a sufficiently strong assumption to allow the solution to proceed. A stronger assumption is therefore

$$w(x,y)=X(x)\sin\frac{n\pi y}{b} \tag{3.5.1}$$

which is a relevant assumption to make for a rectangular plate, simply supported along y = 0, b.

Consider the simply supported rectangular plate, Fig. 3.3. Suppose the pressure loading to be sine varying in each of the x and y directions as indicated.

Then the solution of $\nabla^4 w = p(x, y)/D$ can proceed by *assuming* that w is given by

$$w=A\,.\sin\frac{\pi x}{a}\sin\frac{\pi y}{b}. \tag{3.5.2}$$

The motivation is that the solution sought should be of the separated variables type *and* should satisfy the edge conditions. This latter requirement is that

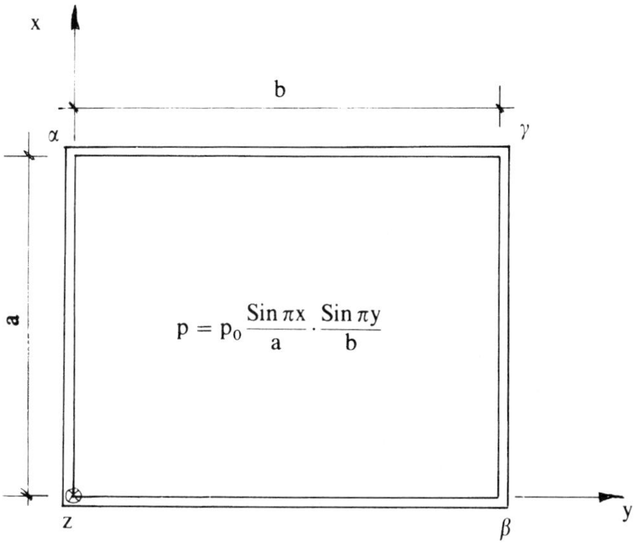

Figure 3.3 Rectangular plate, simple support.

$$0 = w(x, 0) = w(0, y) = M_{yy}(x, 0) = M_{xx}(0, y),$$

and (3.5.3)

$$0 = w(x, b) = w(a, y) = M_{yy}(x, b) = M_{xx}(a, y).$$

These conditions look rather formidable. What they imply is that the simple support conditions must be satisfied. The conditions are the plate equivalent of the ordinary beam simple support conditions, namely that the displacement and the moment *about* the edge line should be zero, all round the boundary.

Thus $w(x, 0) = 0$ is the requirement that the edge displacement along the edge 0α should be zero. The associated moment condition along this edge is that $M_{yy}(x, 0) = 0$, since the simple support condition requires that the plate rest on the edge, supported so as to ensure that $w(x, 0) = 0$ but not restrained rotationally, and hence $M_{yy}(x, 0) = 0$.

Now it can be seen that the choice of an x variation in (3.5.2) of $\sin(\pi x/a)$ is the only one available which will ensure the satisfaction of these edge conditions, for $\sin(\pi x/a)$ goes to zero at $x = 0$, a and hence ensures that $w(0, y) = w(a, y) = 0$.

In addition, along the edge 0β, $\kappa_{yy} = 0$ by virtue of $w(0, y) = 0$, and hence

$$M_{xx}(0,y) = D[\kappa_{xx}(0, y) + \nu\kappa_{yy}(0, y)]$$
$$= D \,.\, w_{,xx}(0,y).$$

Now since w has been assumed to have an x variation of sin ($\pi x/a$), then this is sufficient to ensure that $w,_{xx}(0, y) = 0$ if $w(0, y) = 0$.

Hence it can be confirmed that *all* the edge conditions (3.5.3) are indeed satisfied by the assumed form of w variation (3.5.2). This is the essential importance of the separation of variables form for w, coupled with the individual variations being sine or cosine functions (depending upon where the origin of coordinates is positioned). It should be added that the importance of these features for the simply supported boundary also contributes to the near uselessness of the separation of variables in most other physically interesting edge conditions.

To proceed with the solution, there remains the single task of ensuring that the assumed displacement form (3.5.2), which contains an unknown amplitude, A, is able to satisfy the governing biharmonic equation. Making the substitution of (3.5.2) into (3.2.2), and recalling the particular p(x, y) proposed here, gives

$$D.A\left[\left(\frac{\pi}{a}\right)^4 + 2\left(\frac{\pi}{a}\right)^2\left(\frac{\pi}{b}\right)^2 + \left(\frac{\pi}{b}\right)^4\right]\sin\frac{\pi x}{a}\sin\frac{\pi y}{b} = p_0 \sin\frac{\pi x}{a}\sin\frac{\pi y}{b}.$$

Making a comparison of coefficients it is seen that

$$D.A = \frac{p_0}{\left[\left(\frac{\pi}{a}\right)^2 + \left(\frac{\pi}{b}\right)^2\right]^2}$$

Then

$$w = \left(\frac{ab}{\pi}\right)^4 \cdot \frac{p_0}{(a^2+b^2)^2} \cdot \frac{1}{D} \qquad (3.5.4)$$

is an exact solution to this particular problem.

Some space has been devoted to this rather special example because it displays in a simple way all the essential features of a separation of variables solution. This particular example is of the simplest type, and the solution consists of a single, simple expression. Most examples encountered in practice are likely to be summations of terms derived from superposition of a series of components of the loading expression, which has been expressed as a double Fourier series in order to allow the solution to proceed along the lines adopted above.

The essential features of the process are, first, consideration of the boundary conditions and what form the separation of variables solutions must take in order to satisfy as many of the boundary conditions as possible, and then a formal solution of the governing equation by substitution and comparison of coefficients.

3.6 Further separation of variable features—rectangular plates

In the previous section the simplest solution of all for a rectangular plate was discussed, and the technique of separation of variables was adopted. In this section this same technique will be used, but in a more general context.

Separation of variables usually implies that at least one coordinate variation can be taken to be of sine or cosine type. Indeed, the usefulness of the technique depends upon such a variation in association with a pair of parallel simply supported edges normal to the associated coordinate direction. This is because such a combination can be made to satisfy the boundary conditions on the pair of simply supported edges. Once this is achieved, the problem can be reduced to an ordinary differential equation in the other variable, and this is a much simpler problem than the original one. The process described here is the so called *Levy method* of solution.

Consider the following example—a square plate, simply supported on one parallel pair of edges, fixed on the remaining pair, with a uniform loading (Fig. 3.4).

According to the (Levy) procedure described above, the simply supported edge conditions are satisfied by a variation in the y direction of cos $n\pi y/a$.

Hence

$$w = \sum \phi_n(x) \, . \cos \frac{n\pi y}{a} \tag{3.6.0}$$

is a suitable form for the displacement when n is odd.

It is easy to check that the edge conditions

$$w\left(x, \pm\frac{a}{2}\right) = M_{yy}\left(x, \pm\frac{a}{2}\right) = 0$$

are satisfied by *each* term in the assumed form for w, (3.6.0). Thus the simply supported edge conditions are implied by (3.6.0). It remains to construct the value for ϕ_n (x) such that the fixed edge conditions should also be satisfied.

First express the pressure loading as a Fourier cosine series, thus

$$p_0 = \sum b_n \, . \cos \frac{n\pi y}{a},$$

where

$$b_n = \frac{4p_0}{\pi} \, . \frac{(-)^{\frac{n+3}{2}}}{n} \tag{3.6.1}$$

with n odd.

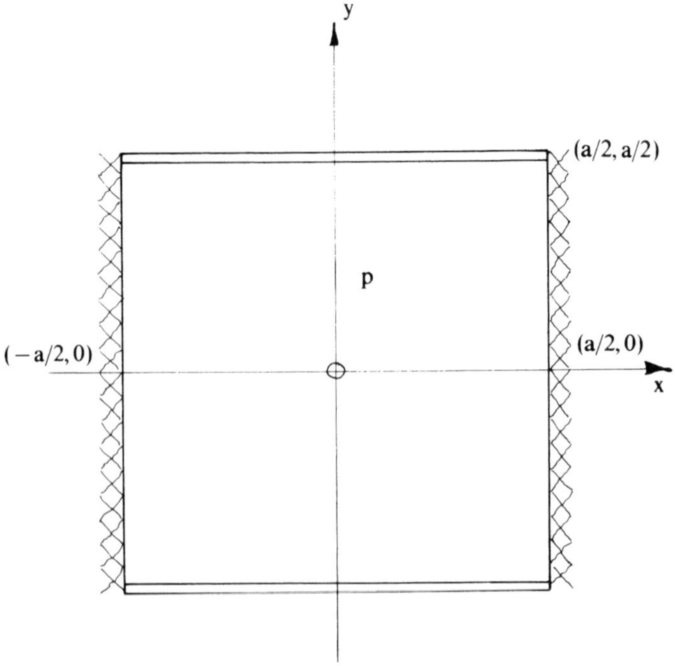

Figure 3.4 Rectangular plate, mixed edge conditions.

Now use (3.6.0) and (3.6.1) in the governing equation (3.2.2) when the following series of ordinary differential equations for ϕ_n is obtained:

$$\phi_n^{IV}(x)+(-2)\left(\frac{n\pi}{a}\right)^2 . \phi_n^{II}(x)+\left(\frac{n\pi}{a}\right)^4 . \phi_n(x)=\frac{4p_0}{\pi D} . \frac{(-)^{\frac{n+3}{2}}}{n} \tag{3.6.2}$$

where n is odd, and the individual equations for ϕ_n (x) have been obtained by separately equating to zero the coefficients of the respective terms in cos ($n\pi y/a$).

The dominant term in the solutions of the problem is likely to be that for n = 1. Hence proceed with the analysis for this term. The equation (3.6.2) is of constant coefficient type, and solutions should be sought of the form $w = Ae^{\alpha x}$. Using this in (3.6.2), and setting the right-hand side to zero for the present, the following equation for α is obtained:

$$\alpha^4 - 2\left(\frac{\pi}{a}\right)^2 . \alpha^2 + \left(\frac{\pi}{a}\right)^4 = 0. \tag{3.6.3}$$

Then

$$\left[\alpha^2 - \left(\frac{\pi}{a}\right)^2\right]^2 = 0, \quad \alpha = \pm\left(\frac{\pi}{a}\right), \text{ repeated.}$$

Hence the basic components in the complementary function solution are sinh $\pi x/a$ and cosh $\pi x/a$. The repeated root indicates that x times these same components are also solutions. Finally, a suitable particular integral for the pressure can be easily constructed by noting that a ϕ_p (x) = const. = K say, gives $\phi_p(x) = 4p_0a^4/\pi^5$.

Hence the general solution of (3.6.2) for n =1 is

$$\phi_1(x) = (A + Bx)\cosh\frac{\pi x}{a} + (C + Ex)\sinh\frac{\pi x}{a} + \frac{4a^4}{\pi^5}\cdot\frac{p_0}{D}. \tag{3.6.4}$$

In the present example, ϕ_1 (x) must be *even* in x, from the symmetry about the x = 0 line.

Hence B = C = 0. There remain the boundary conditions on x = a/2 to be satisfied, then those on x = −a/2 will be satisfied by symmetry.

Hence, at x = a/2

$$\begin{aligned} w\left(\frac{a}{2}, y\right) &= \phi_1\left(\frac{a}{2}\right).\cos\frac{\pi y}{a} \\ &= \left(A.\cosh\frac{\pi}{2} + E.\frac{a}{2}\sinh\frac{\pi}{2} + \frac{4a^4}{\pi^5}\cdot\frac{p_0}{D}\right)\cos\frac{\pi y}{a} \\ &= 0 \end{aligned}$$

and

$$\begin{aligned} w_{,x}\left(\frac{a}{2}, y\right) &= \frac{d\phi_1}{dx}.\cos\frac{\pi y}{a} \\ &= \left(A.\left(\frac{\pi}{a}\right).\sinh\frac{\pi}{2} + E\left[\sinh\frac{\pi}{2} + \frac{\pi}{2}\cosh\frac{\pi}{2}\right]\right).\cos\frac{\pi y}{a} = 0. \end{aligned}$$

Solving

$$E = -1.16\frac{A}{a} \quad \text{and} \quad A = \frac{-3.41}{\pi^5}\cdot\frac{p_0a^4}{D}. \tag{3.6.5}$$

Then

$$(w_{max})_{n=1} = w(0,0) = A + E.0 + \frac{4a^4}{\pi^5}.\frac{p_0}{D}$$

$$= 0.0211\frac{p_0 a^4}{Et^3}, \text{(using } \nu = 0.3). \qquad (3.6.6)$$

The complete solution, using all the odd values of n as implied by the Fourier series (3.6.1), can be shown to give $w_{max} = 0.0209\ p_0a^4/Et^3$.

The bending moments can also now be calculated, and for example,

$$M_{xx}\left(\pm\frac{a}{2},0\right) = D(w,_{xx} + \nu w,_{yy}]_{a/2,0}$$

$$= D\left[\left(\frac{\pi}{a}\right)^2 A\cosh\frac{\pi}{2} + 2E\frac{\pi}{a}.\cosh\frac{\pi}{2} + E\frac{a}{2}\left(\frac{\pi}{a}\right)^2 \sinh\frac{\pi}{2}\right] \qquad (3.6.7)$$

$$= 0.075.p_0a^2, \text{hogging.}$$

This value compares with the complete solution value of

$$M_{xx}\left(\frac{a}{2},0\right) = 0.070.p_0a^2.$$

Exercise: Show that the above values for n = 1 case give $M_{xx}(0, 0) = -0.035p_0a^2$, $M_{yy}(0, 0) = -0.028p_0a^2$, $M_{yy}(0, 0) = -0.028p_0a^2$. The exact values for a uniform load, including all Fourier components for n, are

$$M_{xx}(0,0) = -0.032pa^2, \quad M_{yy}(0,0) = -0.024pa^2.$$

A negative moment indicates a sagging value, tension on underside.

3.7 Solution by finite differences

Thus far only analytical solutions for the plate equation (3.2.2) have been considered. In many situations of practical interest, analytical solutions are not attainable, usually because the boundary conditions do not conform to a type which can be handled by separation of variables methods. Hence numerical solutions must be sought.

A very suitable technique is the *finite difference method.* Here the aim is to evaluate the transverse displacement at a regular grid of points across the plate, rather than construct functional expressions for the displacement. When set up,

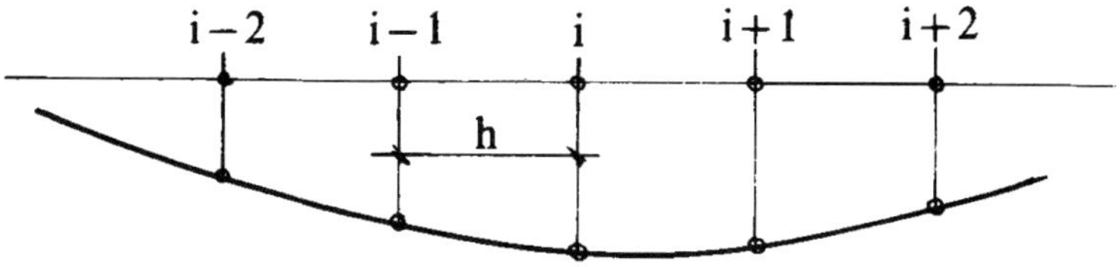

Figure 3.5 Difference grid—one dimension.

the problem becomes an algebraic one of solving a set of simultaneous, algebraic equations for the values of w at the chosen grid of points. Thus the differential equation aspect of the problem is removed by the approximation procedure. The basis of the method is as follows.

Consider first a one-dimensional situation and let there be a regular set of nodal points spaced h apart (Fig. 3.5). For definiteness, Fig. 3.5 can be considered to be a section through a deflected plate. The aim is to relate the values of w at the nodal points $i-2, i-1, i, i+1, i+2$ etc. to the derivatives of w at the central point i.

Now w_{i+1} (x) can be expanded as a Taylor series as follows

$$w_{i+1}(x) = w_i(x) + h\frac{dw_i(x)}{dx} + \frac{h^2}{2!}\frac{d^2w_i(x)}{dx^2} + \cdots$$

Likewise

$$w_{i-1}(x) = w_i(x) - h\frac{dw_i(x)}{dx} + \frac{h^2}{2!}\frac{d^2w_i(x)}{dx^2} - \cdots$$

If h can be considered to be small, say in relation to the total span of the plate, then it should be legitimate to truncate these Taylor series and solve for dw_i/dx (x) in terms of nodal values of w_i. If terms in h^2 and higher powers of h are neglected, then solving for dw_i/dx gives

$$\frac{dw_i(x)}{dx} = \frac{1}{2h}[w_{i+1}(x) - w_{i-1}(x)]. \qquad (3.7.0)$$

This is a typical finite difference expression, where on the left-hand side is a sought derivative, and on the right, the appropriate array of nodal values of w(x).

Proceeding thus, the second derivative can be found, by neglecting terms in h^3 and beyond, to be

$$\frac{d^2w_i(x)}{dx^2} = \frac{1}{h^2}[w_{i+1}(x) + w_{i-1}(x) - 2w_i(x)]. \tag{3.7.1}$$

This array will be represented by the following "module",

$$h^2 \cdot \frac{d^2w_i(x)}{dx^2} = \left[\underset{i-1}{(1)} — \underset{i}{((-2))} — \underset{i+1}{(1)} \right] w$$

where the subscripts will often be omitted, but the meaning will be clear from the context.

Again

$$h^4 \cdot \frac{d^4w_i(x)}{dx^4} = (1) — (-4) — ((6)) — (-4) — (1) \tag{3.7.2}$$

Indeed, the coefficients in the modules can be seen to be given by rows in a Pascal triangle array, with alternate (+) and (−) signs.

For example, beam bending is described by $Dw,_{xxxx}(x) = p$. Use the finite difference method to estimate the deflection for a uniform uniformly loaded, simply supported beam. Suppose h = L/4 is chosen (Fig. 3.6). Then there are 5 nodal points 1−5, and hence 5 w values being sought. But $w_1 = w_5 = 0$, from symmetry and end conditions. Also, from the remaining boundary condition, $M_1 = M_5 = 0$, and when expressed in w terms this gives $w_4 + w_6 - 2w_5 = 0$, or $w_6 = -w_4$. Now the nodes 0, 6 are imaginary, though their notional values will be required.

If the module (3.7.2) is positioned at node 2 and then node 3, with the symmetry requirement $w_4 = w_2$, two algebraic equations for w_2 and w_3 are obtained. Thus

$$\underset{\text{node } 0}{(1)} — \underset{1}{(-4)} — \underset{2}{((6))} — \underset{3}{(-4)} — \underset{4}{(1)}$$

$$= ph^4/D$$

gives

$$6w_2 - w_2 + w_2 - 4w_3 = \frac{ph^4}{D},$$

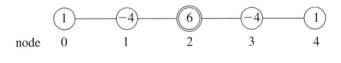

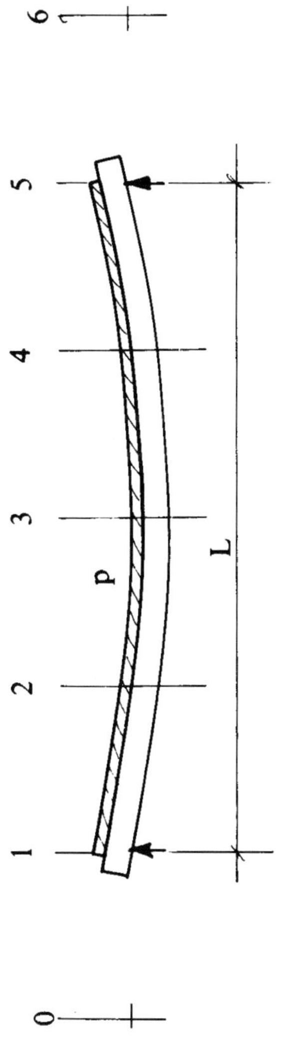

Figure 3.6 Example—difference grid on beam.

and

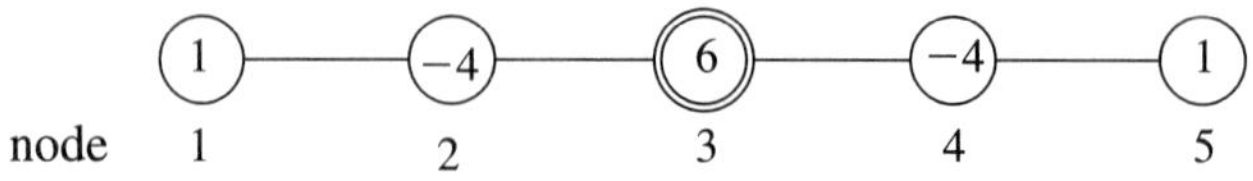

$= ph^4/D$

gives

$$-8w_2 + 6w_3 = \frac{ph^4}{D},$$

or

$$\begin{bmatrix} 6 & -4 \\ -8 & 6 \end{bmatrix} \cdot \begin{bmatrix} w_2 \\ w_3 \end{bmatrix} = \frac{ph^4}{D} \cdot \begin{bmatrix} 1 \\ 1 \end{bmatrix}, \quad \begin{bmatrix} w_2 \\ w_3 \end{bmatrix} = \frac{ph^4}{4D} \begin{bmatrix} 6 & 4 \\ 8 & 6 \end{bmatrix} \cdot \begin{bmatrix} 1 \\ 1 \end{bmatrix}.$$

Hence

$$w_3 = \frac{14}{4} \cdot \frac{p\left(\frac{L}{4}\right)^4}{D} = \frac{7pL^4}{512D} = 0.0136\frac{pL^4}{D}.$$

The exact solution is of course

$$w = \frac{5}{384}\frac{pL^4}{D} = 0.0130pL^4/D.$$

This result is good although only a crude mesh of points was used.

Exercise: Repeat the previous example, but with a central point loading, W, instead of a pressure load. Show that w_{max} = 0.0234 WL^3/D for h = L/4, compared with the exact value of 0.0208 WL^3/D.

Plate problems require that the governing equation (3.2.2) be approximated by a suitable finite difference module. The terms such as $w_{,4x}$, $w_{,4y}$ have been derived from (3.7.2). There remains to construct the module for $w_{,2x2y}$. This can be carried through by formal manipulation of the Taylor series expansion in two independent variables. However, it is preferred here to argue physically with the use of the module for the second derivative (3.7.1).

Consider the array of nodal points (Fig. 3.7). The object is to evaluate $w_{0,2x2y}$. Now

$$h^2 w_{1,2y} = w_5 + w_8 - 2w_1,$$
$$h^2 w_{0,2y} = w_2 + w_4 - 2w_0,$$
$$h^2 w_{3,2y} = w_6 + w_7 - 2w_3,$$

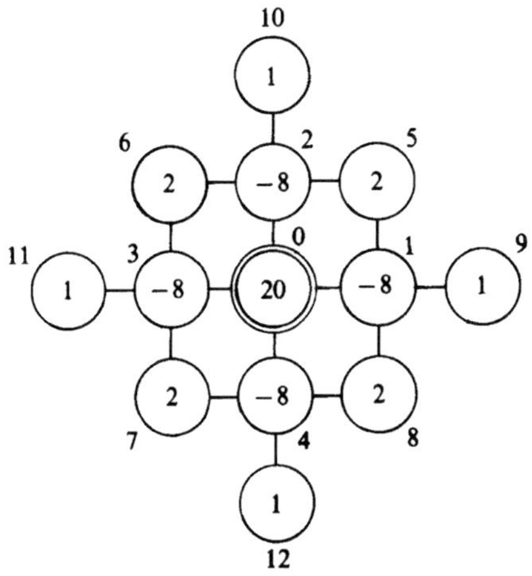

Figure 3.7 Biharmonic difference operator.

and

$$h^2(\ )_{0,2x} = (\ )_1 + (\ )_3 - 2(\ )_0 .$$

Hence

$$h^2(h^2 w_{0,2y})_{,2x} = (w_5 + w_8 - 2w_1) + (w_6 + w_7 - 2w_3) - 2(w_2 + w_4 - 2w_0)$$

Simplify, when

$$h^4 . w_{0,2x2y} = (w_5 + w_6 + w_7 + w_8) - 2(w_1 + w_2 + w_3 + w_4) + 4w_0. \qquad (3.7.3)$$

This information is represented as the module

$$h^4 \frac{d^4 w_0}{dx^2 dy^2} =$$

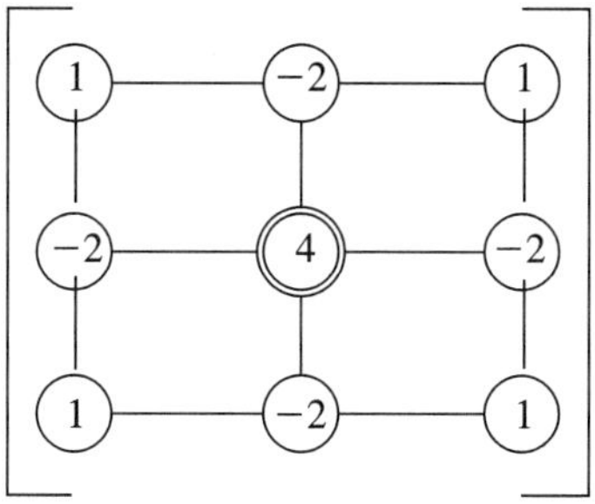

Now collect the three terms in (3.2.2) when the final complete module for the bi-harmonic becomes

$$w_{,4x} + 2w_{,2x2y} + w_{,4y} = \frac{p}{D}$$

or

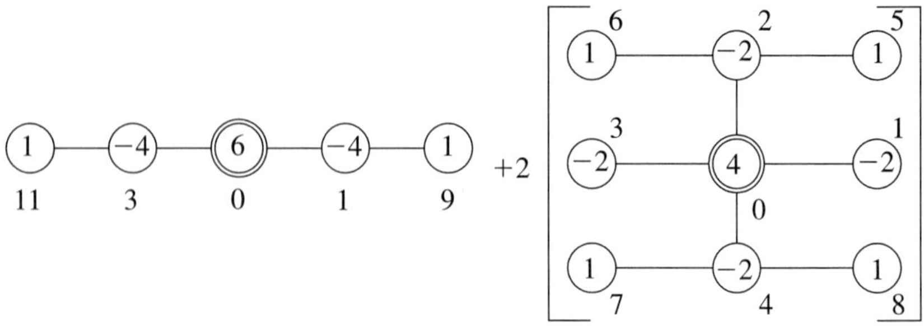

\+

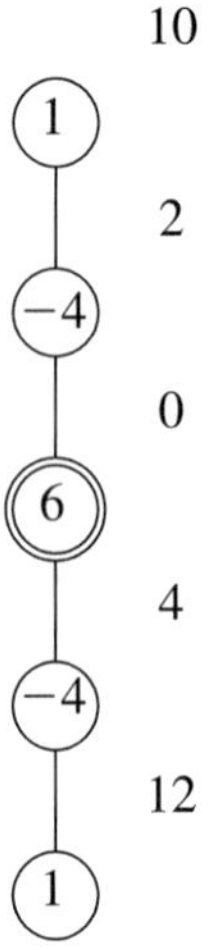

$= ph^4/D$

Namely

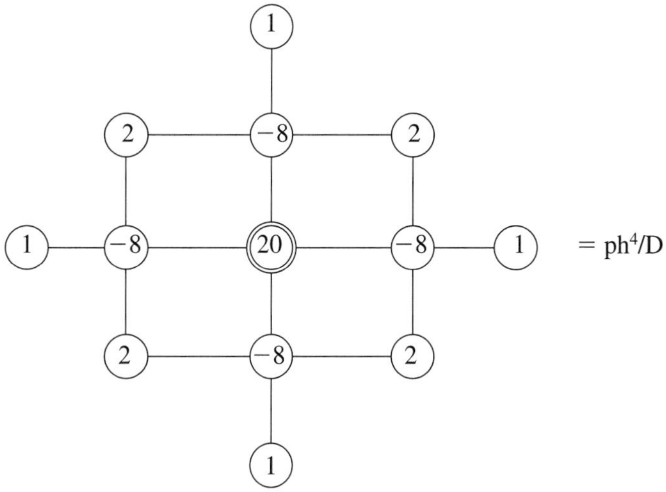

This is the basic plate module, which can be used in exactly the same way as the beam module (3.7.2).

For example, consider a square, simply supported plate (side a) loaded with a uniformly distributed load, p. To begin with, take $h = a/4$. Then, with use of symmetry, there are just three independent unknowns (Fig. 3.8). The module (3.7.4) placed in turn at these three nodal points gives three algebraic equations in the three sought unknown w values.

As in the beam example, so too here, there is a need to identify some fictitious points, shown as 3 and 4 on Fig. 3.8. The simple support conditions then show that the value of w to be associated with w_4 is $-w_1$ and that of w_3 to be $-w_2$.

Hence, with the module centred at 0, 2, 1 the following three equations are obtained:

$$20w_0 + 8w_1 - 32w_2 = \frac{ph^4}{D}$$

$$-8w_0 - 16w_1 + (20+4+1-1)w_2 = \frac{ph^4}{D}$$

$$2w_0 + (20+1+1-1-1)w_1 - 16w_2 = \frac{ph^4}{D},$$

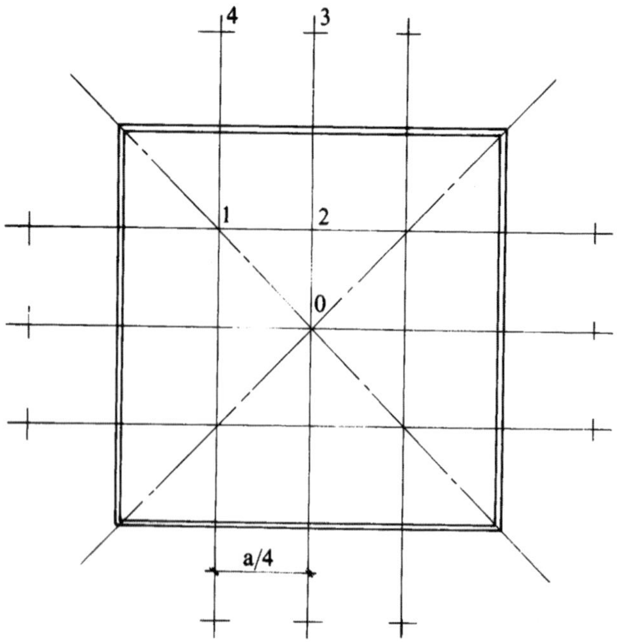

Figure 3.8 Difference grid—simply supported plate.

or

$$\begin{bmatrix} 20 & +8 & -32 \\ 2 & 20 & -16 \\ -8 & -16 & 24 \end{bmatrix} . \begin{bmatrix} w_0 \\ w_1 \\ w_2 \end{bmatrix} = \frac{ph^4}{D} \begin{bmatrix} 1 \\ 1 \\ 1 \end{bmatrix}.$$

Solving, obtain

$$w_0 = \frac{33}{32} \frac{ph^4}{D} = 0.00403 \frac{pa^4}{D}.$$

This compares with the exact value of 0.00406 pa^4/D. The other values are $w_1 = 35/64 . ph^4/D$ and $w_2 = 3/4 . ph^4/D$.

Bending moments can be calculated from these w nodal values. For example,

$$(M_{xx})_0 = D[w,_{xx} + \nu w,_{yy}]_0 = D(1+\nu)w,_{xx} = \frac{D}{h^2}(1+\nu)[2w_2 - 2w_0].$$

and

$$(M_{xx})_0 = -0.0456pa^2 \text{(sagging) (using } \nu = 0.3).$$

This value compares with the exact value of $-0.0479pa^2$.

It must be admitted that the agreement, particularly on bending moment, from use with such a crude grid of points, is better here than can usually be expected.

Exercises: (1) An even cruder solution of this problem is possible, using $h = a/2$. Show that the estimate for w_0 in this case is $0.00390.pa^4/D$, and that the maximum (sagging) moment is given by $-0.0406\ pa^2$ when $\nu = 0.3$.

(2) Use finite differences to solve, approximately, the example dealt with in Section 3.6—square plate simply supported on two and fixed on two other edges. Crudest estimate,

$$h = \frac{a}{2},\ w_{max} = \frac{1}{20}\frac{ph^4}{D} = \frac{1}{320}\frac{pa^4}{D} = 0.0031\frac{pa^4}{D}.$$

For $h = a/4$, show $w_{max} = 0.0024\ pa^4/D$. The exact value is $0.0019\ pa^4/D$.

(3) Repeat the above type of calculation for a clamped-edge square plate under a uniform loading, p. Consider three cases, h=a/2, a/4 and a/6. These give in turn one, three and six unknown nodal values to solve for. Show that the central displacement estimates (w_0 = const. × p . a^4/D) are: 1/24, 328/712 and 411 219/206 792, each to be divided respectively by 2, 4 and 6 raised to the power 4. These data are exact solutions to the respective approximations and can examined as a group to extrapolate to the "best" value by using the "Southwell plot" described below. The values are w_0= 0.0026, 0.0018 and 0.0015 × pa^4/D, respectively, and plotted as described below they give an estimate close to the exact value of 0.0013 × pa^4/D.

The finite difference method when used with small values of h/L, and so generating a large number of unknown nodal values, can provide very accurate solutions. This was the established method before the advent of the computing power now available. Then the solutions were essentially calculated manually using schemes to deal with the large numbers of unknowns. One worker in this field in the period between the two World Wars was R. V. Southwell. In a series of books (see Bibliography, Chapter 7) he described his approach as the "Relaxation" method. This was a one word description for the process devised to solve manually the large sets of algebraic equations being generated. Solving large sets of algebraic equations is now a well catered for task in the computer age and so there should not be a need to avoid large systems. But there may still be occasions when simple manual calculations can throw useful light on a solution.

An example of this sort is to use a sequence of approximate manual solutions, of the type sought in exercise (3) above, to extrapolate to the values we might expect if a much larger number of unknowns had been used. This approach also has its origins in another method put forward by Southwell when he was dealing with the analysis of experimental results from column buckling tests. The method entered the literature as the so-called Southwell plot. A similar approach can be used here to extrapolate to solutions for smaller h/L values.

If we have the approximate solutions for say three even quite large values of h/L as we have here then a suggestion is to make a plot as follows. Define $\Omega \equiv L/h$, and make a plot of Ω/w_0 v Ω, using the available values of w_0. What is expected is a near straight-line plot. The *inverse* slope of this line is then the "best" estimate for w_0.

The reader now has available the apparatus in terms of finite difference modules to set up problems with many more unknowns. The resulting algebraic equations will however require a computer solution.

3.8 Some other aspects of plate theory

Thus far only the simplest types of plate problem have been considered. It is not the intention to delve too deeply into the more difficult aspects of the theory, but some further points can usefully be considered.

The first of these relates to the way in which free edges can be treated. On first encounter, a free edge would appear to be a simple matter to deal with. In beam terms, the presence of a free-end signifies a determinate aspect to a problem. In plates however the situation is quite different. Indeed, the free edge presents the most difficult type of elementary edge condition. The reasons are two-fold.

First there is the circumstance that at a transverse cut in a plate *three* force variables in general appear, a bending moment, a twisting moment and a transverse shear force. If this cut is to be a *free* edge then on intuitive grounds we would expect all three quantities to be zero. However, the *theory* is of *fourth* order only (the biharmonic (3.2.3)) and such a theory is capable of sustaining only *two* boundary conditions. This point will be explained shortly.

The second reason for difficulties presented by free edges is a manipulative one, and arises because the moments and shear forces are respectively second and third derivatives of the transverse displacement. The effect of this is to require boundary conditions to be imposed which are expressed as second and third derivatives of the displacement w. Thinking back to the processes required of a finite difference solution, it will be seen that such a requirement normally means more fictitious points are required, and this increases the problem complexity and hence the total difficulty.

To return to the question of three quantities at an edge and only two permissible conditions. On an examination of the edge Q_x force and the twisting moment M_{xy} on a face x = const. say which is either a supported or a free edge then it is concluded that these two quantities should be grouped to give a composite quantity

$$V_x = Q_x + \frac{\partial M_{xy}}{\partial y}. \tag{3.8.0}$$

This expression results from the observation made in (2.1). There it was noted that M_{xy} can be visualized as a distribution of Q_x type forces forming couples of strength M_{xy}. As M_{xy} varies so there is a near cancellation of the M_{xy} forces at any point seen from left and right. Only $\partial M_{xy}/\partial y$ remains, and hence the above expression (3.8.0) results.

Then at a *supported* edge V_x supplies the vertical edge reaction while at a *free* edge $V_x = 0$ is one boundary condition to be coupled with the second condition $M_{xx} = 0$. Hence there are just *two* conditions.

It is not proposed to explore this question much further except to note that this combining of Q_x and M_{xy} leads to the appearance of concentrated corner forces, typically of value $2M_{xy}$ at a corner in a rectangular plate. The following advice is also offered: in a first look at a plate problem with a free edge, especially if some approximate solution such as a finite difference solution is being sought, then rather than approach the solution in a fully rigorous way, expect that in the neighbourhood of the free edge the plate will tend toward a *cylindrical* shape. By suitable interpretation of this tendency it is often possible to avoid introducing many, or any, fictitious points with a consequent simplification of the problem formulation.

It has been suggested already that most practical elastic plate problems, which are not standard problems, are likely to be solved numerically, since there is so little scope to solve these problems analytically. Hence a method of numerical solution must be decided upon, and today there is a strong likelihood that this choice will come down in favour of a Finite Element approach.

The essential aim of the *Finite Element method* (F.E.M.) is to reduce the continuum plate problem to one which can adequately be described by a *finite* number of parameters. This process is one of approximation and as with all approximation procedures, experience and judgement are needed to fully exploit the potential of the method.

The Finite Difference method described in the previous section is an approximation procedure which is essentially mathematically based. The difference calculus is a branch of classical mathematical analysis, and the only real choice open to the user of the method is the mesh size which is to be adopted.

The finite element method is a physically based approximation procedure, and the user of the method is faced with a large number of decisions in the formulation of the whole finite element calculation. Herein lies the scope for experience and judgement to play a part in achieving an acceptable solution.

In a work such as this, only the briefest of introduction can be given to this method. In essential features the F.E.M. solution uses a stiffness matrix formulation based on an assumed set of element characteristics. There are now many specialized accounts of the method, for example Chapter 5 of J. S. Przemieniecki's Theory of Matrix Structural Analysis (McGraw-Hill, 1968). In real situations it is becoming less likely that individual users will themselves write the necessary computer program to carry through an F. E. M. analysis—it is more likely that commercial software will be used. Then the onus is on the recipient of the results to check the feasibility of what has been obtained, from the point of view of how well the F.E.M. model has been able to simulate the problem in hand. It is in this part of the problem exploration that some of the foregoing methods and results can be employed.

As a sample of the F.E.M. consider a rectangular bending element (Fig. 3.9*a*). Eight parameters, the four corner transverse displacements and the four side curvatures, will be taken to describe the element shape. Alone however these eight quantities cannot define w(x, y) internal to the element. Hence *assume* additionally that the curvatures vary *linearly* away from the edges in each direction. Then it follows that $w_p(x, y)$ at P is given by

$$\begin{aligned} w_p = {} & w_1\left(1-\frac{x}{h}-\frac{y}{k}\right)+w_2\frac{x}{h}+w_4\frac{y}{k}=(w_1-w_2+w_3-w_4)\frac{xy}{hk} \\ & +\frac{\alpha}{2}x(x-h)+\frac{\delta}{2}y(y-k)-\frac{\alpha}{2}\frac{yx}{k}(x-h)+\frac{\beta}{2}\frac{x}{h}y(y-k) \\ & +\frac{\gamma}{2}\frac{y}{k}x(x-h)-\frac{\delta}{2}\frac{x}{h}y(y-k). \end{aligned} \tag{3.8.1}$$

This is a very crude and not very useful shape, but it can serve to illustrate some points of the method. The shape automatically allows element displacement but not slope continuity. For example,

$$\nabla^4 w_p = 0, \tag{3.8.2}$$

showing that the element is incapable of supporting distributed loading. However, suppose the problem of a point loaded rectangular plate. (Fig. 3.9*b*), is to be studied. Four elements with obvious symmetries can span the plate, though in any serious study many more would be used.

From the support conditions it is the case that for a typical element

$$w_2 = w_3 = w_4 = 0 = \beta = \gamma. \tag{3.8.3}$$

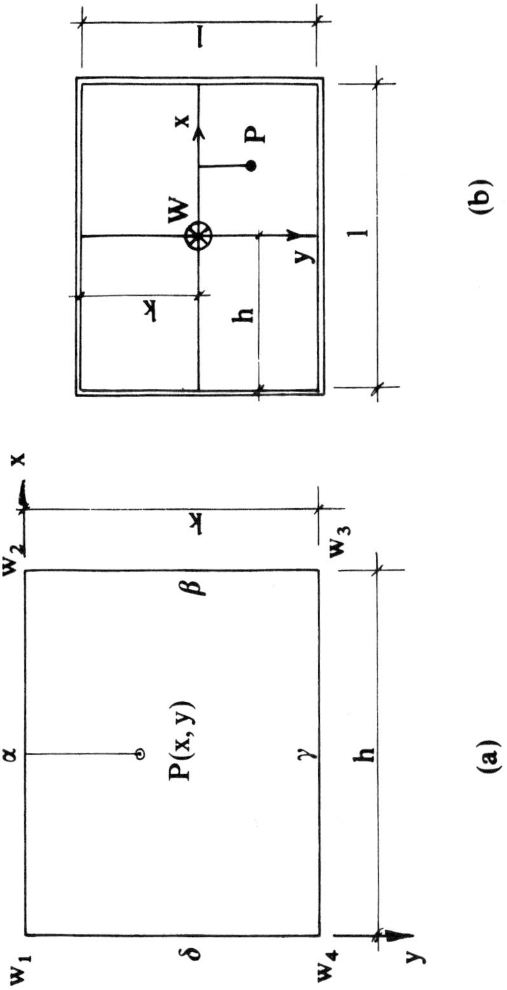

Figure 3.9 Finite elements—element and example.

Then from remarks made earlier (2.1 and 3.8) it is to be expected that point corner forces are generated of value $2M_{xy}$. These summed at the central node will support the point load, W. Thus

$$(M_{xy})_1 = D(w,_{xy})_1 = \frac{Dw_1}{hk} \tag{3.8.4}$$

Hence

$$W = 4x2x\frac{Dw_1}{hk} \quad \text{or} \quad w_1 = \frac{WLl}{32D}. \tag{3.8.5}$$

This is a very poor estimate of w, around three times the accepted value, but serves to show how quickly certain information can be obtained by such approximation procedures.

The remaining parameters α, δ, the element edge curvatures, can be found from the slope continuity condition at the centre to be

$$\alpha = \frac{-2w_1}{h^2}, \quad \delta = \frac{-2w_1}{k^2}. \tag{3.8.6}$$

Now internal force information can be found. For example, for $\nu = 0$,

$$(M_{xx})_1 = Dw,_{xx} = D\alpha = \frac{-W}{4}\frac{l}{L}. \tag{3.8.7}$$

In effect this problem has a single displacement unknown, w_1. When many elements are used matrix methods are called for. Thus suppose the material response to be given by

$$\boldsymbol{\sigma} = \mathbf{De}, \tag{3.8.8}$$

where $\boldsymbol{\sigma}$, **e** are the internal stress and strain vectors, and **D** is the symmetric material response matrix—this stage corresponds to (3.1.3). Generally $\boldsymbol{\sigma}$ and **e** will vary from point to point. The vector of nodal displacements **d** and the strains **e** will be related through some chosen element properties, such as (3.8.1), and can be written symbolically as

$$\mathbf{e} = \mathbf{B} \,.\, \mathbf{d}. \tag{3.8.9}$$

Hence if the external forces **p** relate to the internal actions $\boldsymbol{\sigma}$ through a contragredient equilibrium relation of the type

$$\mathbf{p} = \int \mathbf{B}^T . \boldsymbol{\sigma} \, dV \tag{3.8.10}$$

where $(\ )^T$ is the matrix transpose and V is the volume of the element, then

$$\mathbf{p} = \int \mathbf{B}^T . \boldsymbol{\sigma}\, dV = \int \mathbf{B}^T . \mathbf{D} . \mathbf{e}\, dV = \int \mathbf{B}^T . \mathbf{D} . \mathbf{Bd} . dV$$
$$= \int \mathbf{B}^T . \mathbf{D} . \mathbf{B}\, dV . \mathbf{d} = \mathbf{K} . \mathbf{d}. \tag{3.8.11}$$

Here $\mathbf{K} = \int \mathbf{B}^T . \mathbf{D} . \mathbf{B}\, dV$ is the element stiffness matrix. By suitable assembly of these element stiffness matrices, a complete F.E.M. problem can be formulated and solved for the nodal displacements, **d**. Thereafter the internal force information is obtained by use of (3.8.8, 9).

Solutions must be carefully scrutinized to ensure that they adequately solve the problem in hand. Looked at from a different point of view, every problem solved provides the exact solution to *some* problem, but usually not the one aimed for.

3.9 Stability of plates

Thus far the formulation of the elastic plate problem has been centred on the bi-harmonic equation (3.2.3). In this section it is proposed to supplement this equation with the necessary terms which will enable the *stability of equilibrium* of the elastic plate to be studied.

The essential differences between the plates studied in this section and those earlier in the chapter are, firstly, the equilibrium here is written down with explicit regard for the equilibrium being required in the *deformed* state of the plate and, secondly, that the typical loading which the plate here must withstand is forces of N_{xx}, membrane type, which act in the original plane of the plate.

The plate is still primarily being *bent*, but the bending is being brought about by in-plane forces acting on the transverse displacements which develop when the plate becomes unstable.

As has been remarked, the loading of interest when studying plate stability is from in-plane rather than out-of-plane forces. These forces of type N_{xx} (Fig. 3.10) then can be seen to exert moments of the type $N_{xx}\, dw$ between the ends of a plate element. Hence the moment equilibrium equation about an axis parallel to the y axis becomes

$$M_{xx,x} + M_{xy,y} + N_{xx} w,_x = Q_x. \tag{3.9.0}$$

The N_{xx} force is transmitted unchanged across the plate, because of the absence of other in-plane forces. Hence $N_{xx,x} = 0$.

Now when the single moment equilibrium equation is formed by eliminating the Q_i, then (3.9.0) contributes a term of the form $N_{xx} . w,_{xx}$. Finally, when the $M_{\alpha\beta}$

values are replaced by the curvature expressions, but expressed in terms of the transverse displacement, then the governing equation for the elastic plate being compressed in one direction is found to be

$$D\nabla^4 w + N_{xx}\, w,_{xx} = 0. \tag{3.9.1}$$

The plus sign in this equation is physically important and indicates that the equation describes an instability phenomenon. If a minus had been present, then no instabilities occur and physically the plate would be under stretching rather than compression forces.

Consider now a solution for (3.9.1), when the physical arrangement of plate and support is as in Fig. 3.10. Assume a separation of variables in *Navier* form of a pair of sine type variations, one in each axis direction.

Hence

$$w(x, y) = C_{mn} \cdot \sin\frac{m\pi x}{a} \sin\frac{n\pi y}{b}. \tag{3.9.2}$$

This choice has the virtue that all the edge (boundary) conditions are automatically satisfied. There remains only the determination of the C_{mn} in order to satisfy the governing equation (3.9.1). Substituting (3.9.2) into (3.9.1) produces

$$C_{mn}\left[D\pi^4\left\{\left(\frac{m}{a}\right)^2 + \left(\frac{n}{b}\right)^2\right\}^2 - N_{xx}\left(\frac{m\pi}{a}\right)^2\right] = 0.$$

The conclusion to be drawn is that either $C_{mn} \equiv 0$, in which case the plate is stable because $w \equiv 0$, or

$$N_{xx} = D\left(\frac{\pi}{a}\right)^2 \cdot \left(\frac{m^2 + n^2\left(\frac{a}{b}\right)^2}{m}\right). \tag{3.9.3}$$

If this is so, then $C_{mn} \neq 0$ is possible, and this would indicate that $w \neq 0$, hence the plate has become unstable.

The parameters m, n must now be chosen to minimize N_{xx}. Suppose $a/b = 1$, then N_{xx} is minimum for $m = n = 1$. For $a/b = 2$, the minimum N_{xx} occurs for $m = 2$, $n = 1$. But in either case the minimum N_{xx} is given by

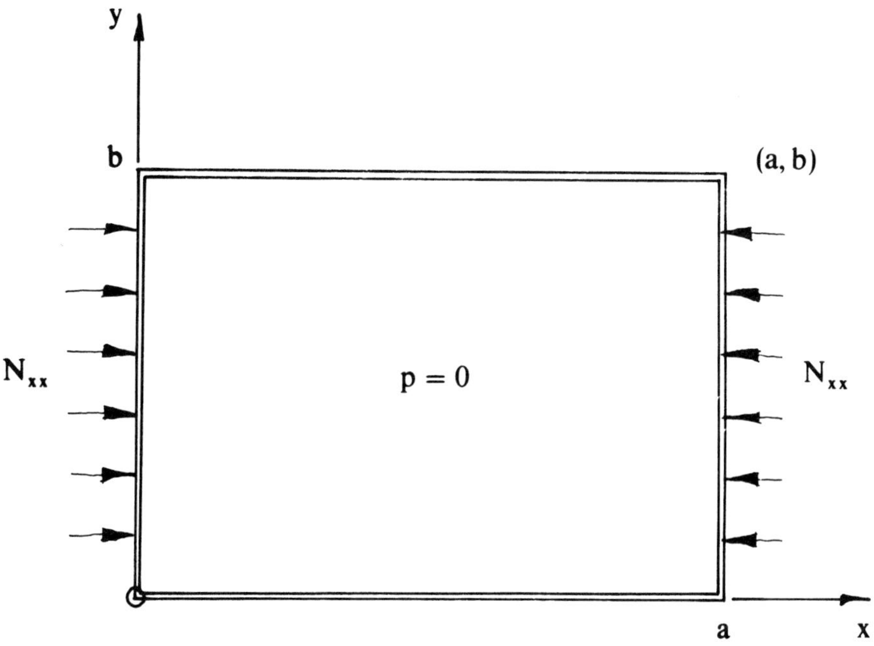

Figure 3.10 Plate stability—uniaxial compression.

$$(N_{xx})_{min} = \frac{4\pi^2 D}{b^2}, \qquad b \leqslant a. \tag{3.9.4}$$

This is a very characteristic form, and strongly resembles the well-known Euler Strut Formula.

It is easy to show that the value (3.9.4) is repeated for all integral m, and n = 1, namely for all integral a/b. But what for a/b not integral? If a plot of N vs. a/b is made by selecting n = 1 and, in turn, integral values at m, and then plotting (3.9.3) as a function of N and a/b, a series of curves will be obtained which all share the common minimum at integral a/b, see Fig. 3.11. From the figure it can be seen that $(N_{xx})_{min}$ for non-integral a/b is greater than for integral a/b, though not much greater. Physically what is occurring is a panel buckling which for a/b integral means square panels of side b(<a). For non-integral a/b square panels cannot be accommodated, and the waves in the long direction change to allow a whole number of half waves to span the whole plate.

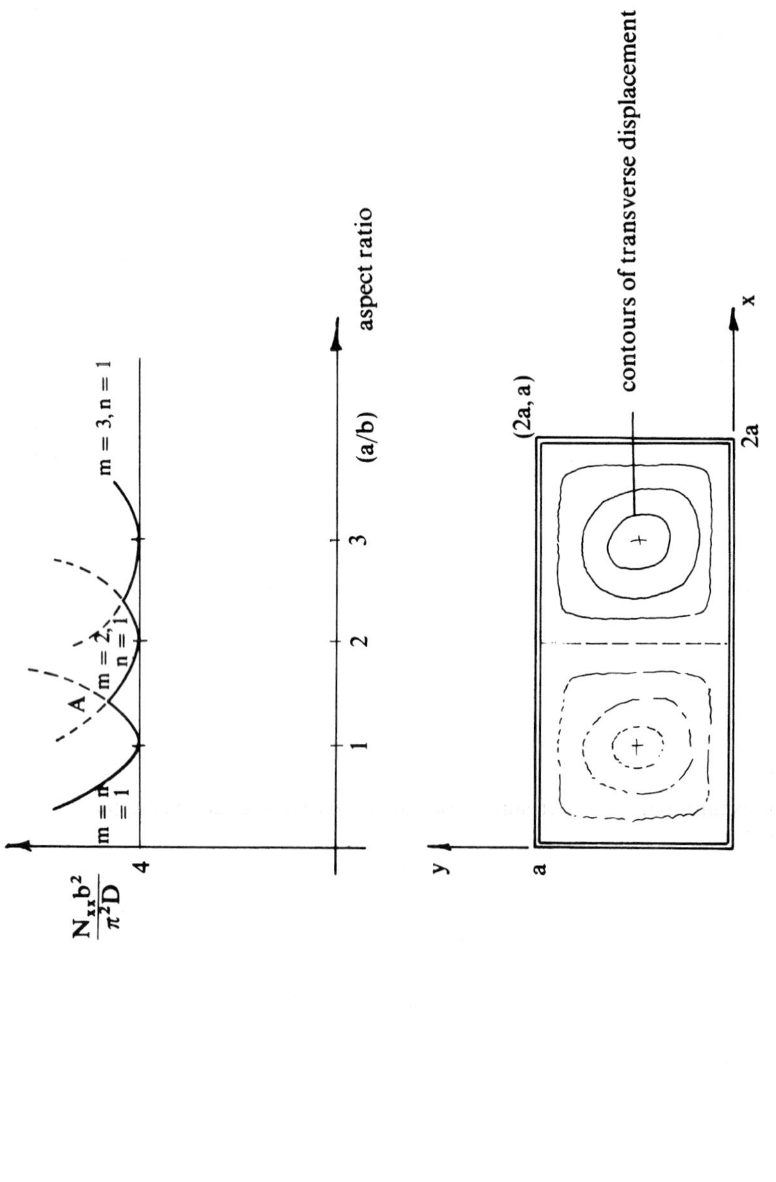

Figure 3.11 Plate stability—typical results.

For safety's sake, and remembering that an instability is being studied, the base values of $N_{xx} = 4\pi^2 D/b^2$ should probably be adopted for all side ratios.

Exercise: Show that the coordinates of point A on Fig. 3.11 are 4.5, $\sqrt{2}$.

3.10 Further exercises

(1) A topic we shall study in more detail in the next chapter relates to non-dimensional numbers that can characterize features of plate behaviour. An example is what we shall term a Stiffness Number, S, defined to be the maximum displacement, w_m, times the plate stiffness, D, divided by the pressure loading, p, times the square of the plate area, A. Thus

$$S \equiv w_m D/(p A^2)$$

First check that this is a non-dimensional combination. Here the dimensions are respectively, length × force × length / force × length^{-2} × length^4. Namely zero dimensions. Show for a circular, pressure loaded simply supported plate that this number, S, has the value 0.00792. One use of this number is to note that there is some reason to believe that this is the largest value this number can take for all simply supported pressure loaded plates of the same area as the circle. The plates may be any shape but cannot have holes or changes of stiffness from place to place. For the square-shaped plate the value is 0.00406. As the shape becomes more elongated so the value of S reduces. What we are saying above is what is termed a *conjecture*. We have not proved that S for the circular shape is greatest. The observation is that the circle has the feature that parts of it, like the centre, are further away from the boundary support than for any other equal-area shape. For this and other reasons we think about how to characterize the very symmetrical and well documented circular shape as compared with all others, and then propose this conjectured property.

(2) For fixed-edge plates the circular plate is five times stiffer than the square-shaped plate of equal area. Thus S = 0.00792/5 = 0.00158. Then follows the further conjecture that: all other shapes of equal area fixed-edge elastic plates have smaller S values.

There are many other possible such relationships. Try coming up with one of your own!

(3) The shape of the plate is clearly important. In practical use of plates in buildings for example, the most common shapes are rectangles. If we restrict our comparison to rectangles only then the square shape now plays the role that the circle played earlier. The S-conjecture then becomes that S for the square is larger than for any other equal area four sided shape. Do you agree and can you demonstrate that the square shape has the largest S value?

3.11 Conclusions

In this chapter the elastic plate problem has been formulated and solved in some very simple cases, and both analytical and numerical solutions have been outlined. The first steps towards more comprehensive solutions have been aimed at. Such solutions will usually involve computer solution of the algebraic equations which are produced, particularly if (say) a finite difference method is adopted. More detailed treatments of this class of plate problem can be found in the works listed in Chapter 7.

4 Plastic plates

4.0 Introduction

As remarked in Chapter 3, a plate near to collapse is likely to be in a plastic condition, whereas a plate at working load is likely to be in an elastic state. Compared with elastic plates, plastic response in plates is more varied, since it depends upon the nature of the plate material as to quite what response is to be expected.

In the first portion of this chapter solid metal plates, typically steel plates, will be studied. Later sections will deal with the response of reinforced concrete plates, but always in the near collapse, maximum load carrying capacity state.

Most plates at most stages of the loading cycle up to collapse have an important feature in common. This is the manner in which the strains are distributed through the plate thickness—the so-called *Kirchhoff hypothesis*. In Chapter 3 this hypothesis is described in some detail. Here it is noted that this same *plane sections* behaviour is found to be a reasonable working hypothesis for plastic plates also. Thus the direct strains parallel to the plate surfaces are proportional to the distance from the middle surface at the section concerned.

I SOLID METAL PLATES

4.1 Yield criteria

In order to describe plate behaviour in the plastic condition, the material response must be defined. This is the role of the *yield criterion*. Consider a simple beam being bent, first of all elastically and then finally plastically (see Fig. 4.0). Initially there is a proportional, recoverable response along the portion OA of the moment/curvature plot. Then the response ceases to be elastic (A $\rightarrow$ B) and finally a plastic straining at constant moment (B $\rightarrow$ C) occurs. For a number of practical cases this full response can be approximated by the rigid-plastic response curve ODC. Such an idealization suggests that the elastic curvatures are *small* when compared with the final curvatures and hence can be neglected compared with the plastic curvatures.

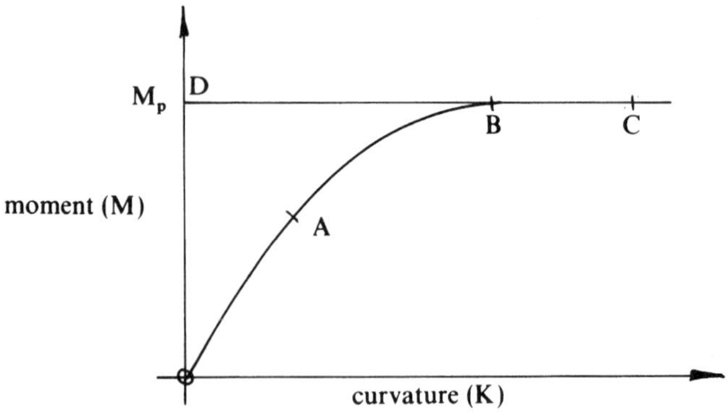

Figure 4.0 Moment-curvature relations—beam element.

This material response data can be presented in another manner. For the beam described above the material response can be thought of as

$$0<|M|<M_p, \tag{4.1.0}$$

that is, M is constrained to be numerically less than M_p, since the beam is unable to sustain a moment greater than M_p. The expression (4.1.0) is termed the yield criterion for the beam.

In the case of the plate, in general at every site in the plate there will exist a field of bending moments—in particular there will be two principal bending moments at each point. The yield criterion is most easily expressed in terms of principal moments.

Consider the yield is criterion (Fig. 4.1). This is a principal moment plot, the so-called "square yield criterion", which is interpreted as meaning that either principal moment can rise to the one-dimensional full plastic value in either sense before yield is reached, and beyond which further moment cannot be sustained. There is some experimental evidence to suggest this is a reasonable criterion. In addition it is the simplest possible response for yielding in two dimensions.

The yield locus (Fig. 4.2) is the *Tresca Locus*, named after an early worker in this field. This locus is built up from the experimental observation that certain ductile materials begin to flow when the maximum shear stress reaches a critical value. As can be seen, the Square and Tresca criteria are essentially identical for moments of the same sign, but different when the moments are of opposing signs.

In either case the interpretation is that any combination of principal moments which plot to a point in the interior of the figures 4.1, 4.2 will be sustained without yield. If the combination plots to a point on the boundary of the figures, then

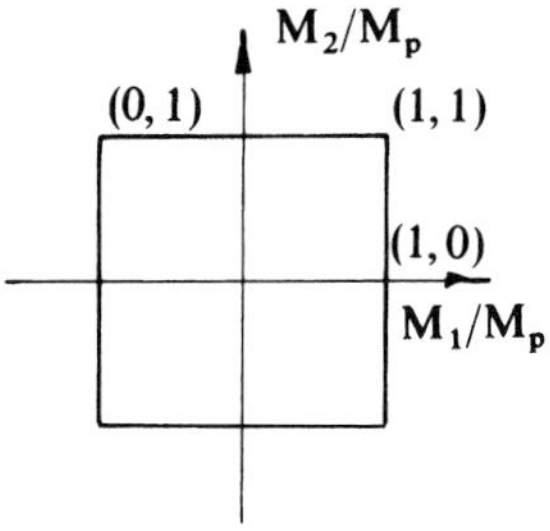

Figure 4.1 Square yield criterion.

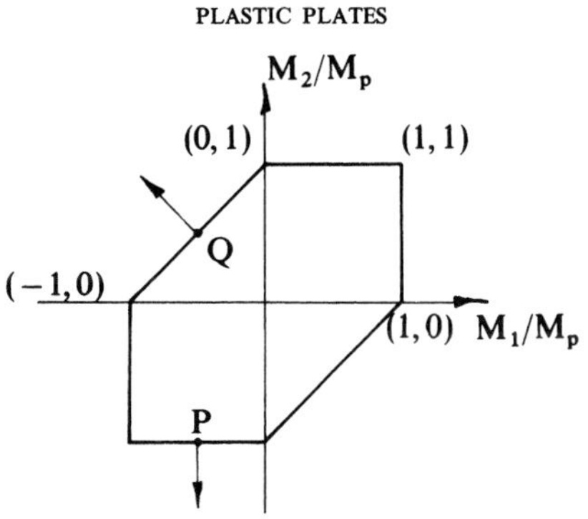

Figure 4.2 Tresca yield criterion.

yield is possible, while moment combinations which lie outside of the figures are unattainable. The two yield criteria so introduced are relevant for two-dimensional moment fields such as those encountered in plates being bent by transverse loads. Thought of at the level of an element of material, the square yield locus is an expression of the principal stress failure theory whereby each principal stress in the plane of the plate can rise to yield without influence from the other. By considering the plate to consist of a series of laminae, all of which on one side of the neutral (central) lamina are stressed in one sense, say compression, while all those on the other are in tension, and integrating through the plate thickness after the manner of 2.1, so the moment resultants are obtained. If the yield stress is f_y and the plate thickness t, then the moment resultant/unit plan length is given by $M = f_y \, . \, t^2/4$. Hence $t^2/4$ is the per unit length, plastic section modulus.

In practical cases the square yield criterion will be used most extensively in the study of non-homogeneous reinforced concrete plates. (See Section II).

The Tresca criterion (Fig. 4.2) is more appropriate to solid homogeneous metal plates. If there is a need to consider the moment resultant in terms of the material yield stress, then the expression derived above ($M = f_y \cdot t^2 / 4$) applies here too. For yielding in the quadrants of Fig. 4.2 where the principal moments are of like sign, the larger principal stress coupled with the zero (or near zero) stress through the plate thickness produces yield by sliding along planes which are inclined at $\pi/4$ to the plate surface and intersect the plate surface along lines which are *parallel* to the *larger* principal moment vector.

By way of contrast, in those quadrants at Fig. 4.2 where the moments are of opposite sign, the (supposed) laminae of plate material are being *compressed* in one principal stress direction but *extended* in the other (Fig. 4.3). Now the maximum principal of stress *difference* is not between the in-plane yield stress and the zero normal stress through the plate thickness, but is the maximum stress difference of $2f_y$ between yield in tension and compression on orthogonal axes. The yielding now occurs on planes which are *normal* to the plate surface, namely planes directed through the plate thickness and equally inclined to the principal moment axes. This change in the planes of yielding has consequences for the type of deformation possible, and will be returned to below (4.3) where the question of curvature change and the Normality Rule are discussed.

4.2 The bound theorems

These theorems will be described and justified as physical consequences of the plastic material properties, and equilibrium. The importance of these theorems derives from the difficulty experienced in fully solving many problems of potential interest. Instead, a restricted formulation of the problem, chosen because it can be solved, will form the basis of the solution, and then the bound theorems will be invoked to indicate whether the real solution implies a greater or lesser collapse load, for example.

In any complete problem there are essentially three component parts of the complete solution. On the one hand *equilibrium* must be achieved—in this case equilibrium at the limit of load-carrying capacity of the plate. Secondly, *yield* will have occurred at certain points or in certain regions of the plate; just where yielding has occurred must be investigated. Additionally the yield criterion must not anywhere be violated. Thirdly, the yield pattern must be such as to constitute a *mechanism* in the plate. The concept of a mechanism is geometrical and is a direct extension into two dimensions of the concept of a mechanism in a beam or framework. The basic requirement of a mechanism is that it should consist of patterns of lines, usually straight lines, seen in plan on the plate, *about* which the

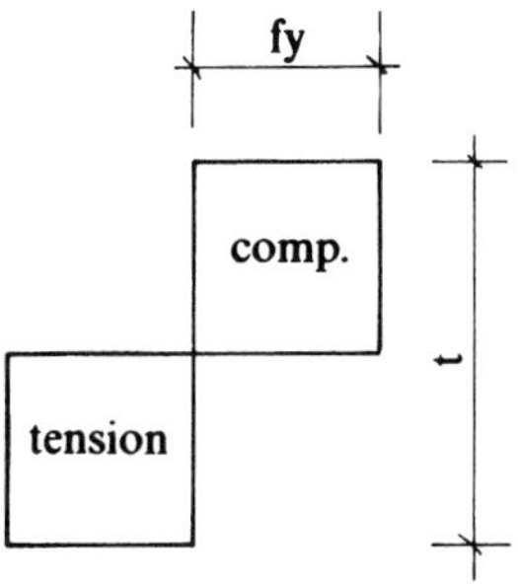

Figure 4.3 Full plastic moments.

adjacent rigid regions of plate can rotate relative to one another. These lines of rotation are termed *yield lines* and must satisfy certain criteria if the whole pattern of lines is to form a mechanism.

For example, consider a square plate supported all round the perimeter (Fig. 4.4). Now if yield lines were to form across the two diagonals, then a *small* rotation of each quarter plate about the supported edge will cause points in the plate interior to move transverse to the plate surface and will produce a relative rotation between quarter plates to develop along the diagonals. The originally plane square plate will deflect into a square base (shallow) pyramid. The rotations must be considered small, and the deflections small compared with the plate span and thickness. With these restrictions, the plate segments will not tend to separate across the yield lines. This is an example of a simple mechanism and will be called a mechanism with *one* degree of freedom.

The one degree of freedom can be taken to be Δ (Fig. 4.4), the central (maximum) displacement, and indicates that *one* displacement, say Δ, can be chosen arbitrarily, whereupon the deflections of all other points in the plate middle surface follow for deformation of the plane middle surface into the pyramid shape.

A typical *two* degree of freedom mechanism might be a pyramid within a pyramid (Fig. 4.5). Now *two* typical displacements, say Δ and δ, are required to determine the final displacements at all points.

In principle, mechanisms with very many degrees of freedom might occur in practice. However, in modelling this behaviour, minimum numbers of degrees of freedom will be assumed in order to keep the analysis as simple as possible, and this number will usually be restricted to one.

It has been said that three aspects must be simultaneously satisfied if the plastic problem is to be solved completely: namely equilibrium, yield and mechanism. The bound theorems relate to situations in which the first *together with one or other of the remaining* two conditions are satisfied in a problem.

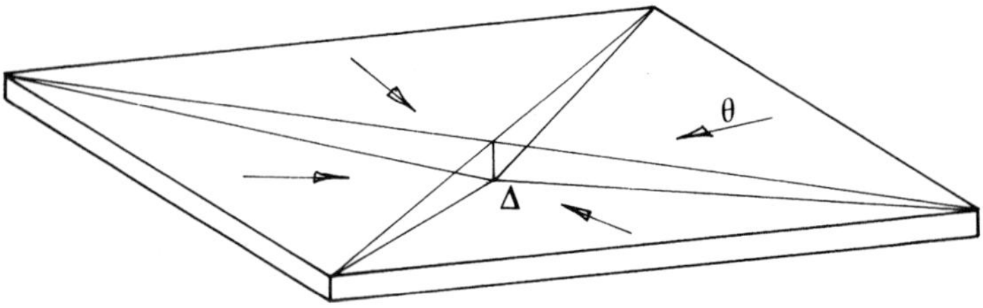

Figure 4.4 Plate Mechanism, one degree of freedom.

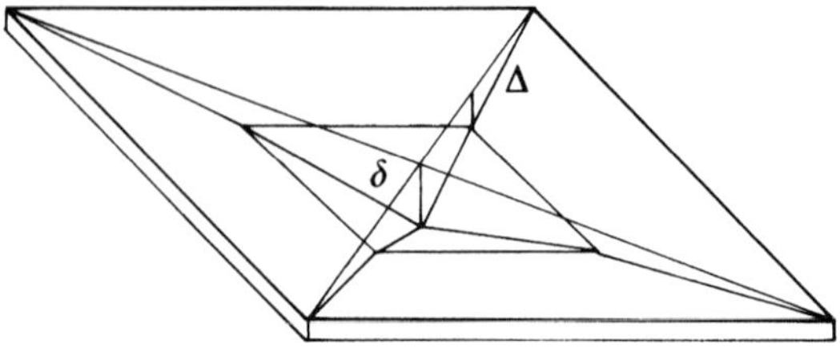

Figure 4.5 Plate mechanism, two degrees of freedom.

Consider the case of a transversely loaded plate when equilibrium is satisfied and the yield criterion is not violated, but there is no guarantee that a mechanism has been achieved. What can be said about the solution for the load obtained in this case? In this case the *Lower Bound (Safe) theorem* will be invoked to say that the *load* (p_L), which is calculated on the basis that equilibrium is satisfied everywhere and that the moments at every point combine to at most achieve yield but do not violate yield anywhere, will be *less than* or at most *equal to* the true collapse load of the plate. The reason is that there is no guarantee that sufficient yield has occurred to form a mechanism, and hence the collapse load has not yet been reached.

On the other hand, suppose equilibrium is achieved and a mechanism is *imposed*, but no direct use of the yield criterion is made, then the *Upper Bound (Unsafe) theorem* says that the load (p_u) calculated this way is *greater* than or is perhaps *equal* to the true collapse load.

Hence we have that the true collapse load (p_c) is bounded above and below by p_L and p_u, i.e.

$$p_L \leq p_c \leq p_u. \tag{4.2.0}$$

It is not proposed to offer formal proofs of these results. Instead, as opportunity arises, the correctness of (4.2.0) will be reinforced by examples.

In case of true design, where the loads are specified and a value for the plate bending strength is required, then the terms "upper" and "lower" must be exchanged. For such reasons, the terms "safe" and "unsafe" are really more appropriate, though "upper" and "lower" are found in the literature more regularly than "unsafe" and "safe".

4.3 The normality rule

The yield criterion is a closed convex figure in moment space (that is moment coordinates). The purpose served by the criterion is to specify what combinations of bending moment, acting at a point in the plate, will cause yield. In addition to this moment information there is also some deformation information implied. This derives from the so-called *normality rule*.

Consider a point such as P in Fig. 4.2. The M_2 moment is at yield while the M_1 moment is not. The outward normal drawn to the yield surface at P points in a direction parallel to the M_2 axis. The normality rule then states that only curvature changes in this direction are possible at point P. In this case then the only possible curvature change is one associated with the yield moment $M_2 = M_p$ at P.

Consider Q (Fig. 4.2). Here, because of the role of the maximum stress (hence moment) difference noted earlier, the outward normal now points in a direction equally inclined to the moment axes. The interpretation now is that the increments of curvature change in the 2 and negative 1 directions must be equal at point Q.

Though very often the information provided by the normality rule will not be used fully, it nevertheless provides an important link between the moment information of the yield criterion and the deformation information on which the construction of mechanisms depends. Examples later will illustrate the use of the normality rule.

4.4 Circular plates—square yield locus

The most satisfactory way of illustrating the use of yield criteria, the bound theorems and the normality rule is by examples. To begin with consider the circular plate, subject to a uniform pressure loading, and suppose the plate to be clamped around the edge.

When thinking about plastic plate problems in particular, it is well worthwhile to visualize such things as the shape into which the plate might deflect. Usually the simplest possible shape, consistent with the type of support the plate must conform to, is the one to investigate, at least for a start on the problem. Thereafter

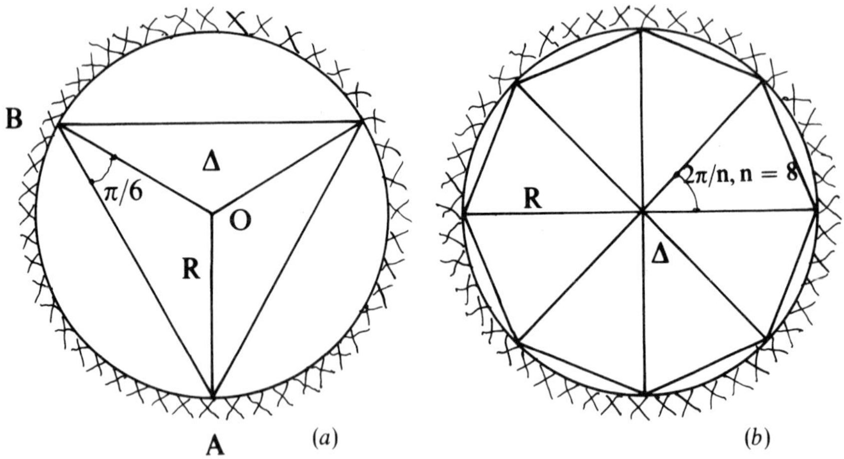

Figure 4.6 Collapse of circular plate. (*a*) n = 3; (*b*) n = n.

related or more complicated shapes often are suggested by what has gone before. Most frequently, a work equation formulation of equilibrium will be used, rather than consideration of the differential equations of equilibrium. For this reason, too, the careful visualization of the deflected shape of the plate is very useful in formulating the relevant terms in the work equation.

In the present case of the circular plate, the most obvious mode of deformation to consider is one whereby the originally plane plate deforms into a pyramid shape. The simplest shape of this type is the triangular-based pyramid (Fig. 4.6*a*). Once Δ, the central displacement, is specified, then the shape is determined. In particular, the angles formed between adjacent plane (deformed) faces of the deflected pyramid shape are determined. Thus the greatest slope on any pyramid face is 2Δ/R. But slope is a *vector* quantity, and hence the resolved part of this slope normal to the hinge line OA is

$$\frac{2\Delta}{R} \cdot \cos \pi / 6 = \frac{\sqrt{3}\Delta}{R}. \tag{4.4.0}$$

Hence the total angle of rotation between the adjacent pyramid faces is $2\sqrt{3}\Delta/R$. This is the relative rotation across a radial hinge line when n = 3. Across an edge hinge line such as AB, the relative rotation is the earlier calculated maximum slope of 2Δ/R.

Consider now the material response. This is described by the yield criterion. The description “plastic” does not fully determine the material response, and the precise yield conditions must be spelled out. For simplicity here it will be assumed that the yield locus is as given in Fig. 4.1—this is the "square" yield locus of earlier and is a good approximation to certain physical material responses, notably reinforced concrete elements with equal reinforcement in each direction and in each face of the plate. But it is also a suitable simple yield condition for steel plates. Here a positive moment will be assumed hogging, and negative, sagging. $M_{1,2}$ refer to the *principal* moments at a given point in the plate. Thus from point to point, the *orientation* of these moments may vary. Such information is not explicitly provided by the yield locus. The non-dimensionalizing quantity, M_p, is the fully-plastic moment, analogous to the uniaxial beam fully plastic moment (see 4.1). In the present example, the n = 3 case, there will be moments in general generated throughout the plate when the pressure loading, p, rises to the collapse value, p_c.

To solve the problem completely, a solution for the internal moments must be constructed in which equilibrium is achieved at all points, yield is nowhere violated and yielding occurs at sufficient points for a yield mechanism to form. To satisfy all these conditions simultaneously is usually not possible. An approximation is sought, and the most fruitful approach is to invoke the bound theorems, and especially the upper bound (unsafe) theorem, assuming the collapse pressure (p_c) is being sought for a given strength (M_p) of plate. Now the search is relaxed to consider just the equilibrium and mechanism conditions. In the present case it will now be assumed that the n = 3 mechanism or mode of deformation will occur, and hence M_p will be *assumed* to act along all the yield lines, here typically OA and AB. Then the internal work absorbed by the M_p acting while the relevant angle change across a yield line develops must just balance the decrease in potential energy of the load, p. Expressed in symbols, this requires that

$$
\begin{aligned}
p\times\tfrac{1}{3}.\tfrac{1}{2}.\left(\sqrt{3}R\right)^2.\sin\pi/3.\Delta &= \text{decrease in P.E. of loading}\\
&= p\times\text{volume swept out by plate as it deflects}\\
&= \text{internal moment work} = \Sigma|M_p|\,.\,|\theta|\,.\,L,\\
&= Mp\left[3\cdot R\cdot\frac{2\sqrt{3}\Delta}{R}+3\sqrt{3}R\cdot\frac{2\Delta}{R}\right].
\end{aligned}
$$

Solving for p,

$$p_3 = 48\frac{M_p}{R^2}. \tag{4.4.1}$$

This *estimate* of the true collapse load, p_c, is, according to the upper bound theorem, either equal to, or greater than, p_c. To investigate the solution further

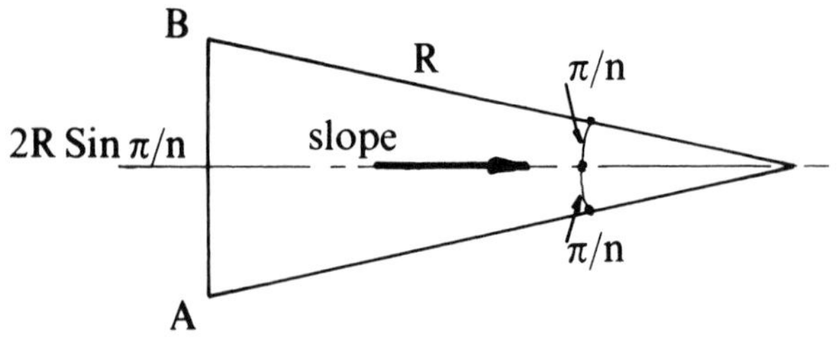

Figure 4.7 Circular plate, typical segment.

suppose the pyramid shape of the deformation is based on a regular “n” sided polygonal shape, rather than just the 3 sided shape.

Now, the greatest slope toward the centre is $\Delta/R \cos \pi/n$ and this, resolved normal to the sagging (radial) yield line, is $\Delta/R \cos \pi/n \,.\, \sin \pi/n$. The work equation becomes

$$p \,.\, \frac{1}{3} n \left(\frac{1}{2} R^2 \sin \frac{2\pi}{n} \right) = M_p \left[n \,.\, \left(2R \sin \frac{2\pi}{n} \right) . \frac{\Delta}{R \cos \pi/n} + n \,.\, \frac{2\Delta}{R} \,.\, \tan \frac{\pi}{n} \,.\, R \right]$$

or

$$p_n = \frac{12M_p}{R^2} \sec^2 \frac{\pi}{n}. \qquad (4.4.2)$$

The minimum value of this expression is for $n = \infty$ when

$$p_c = \frac{12M_p}{R^2}. \qquad (4.4.3)$$

Hence the most likely mode shape is the circular cone. This is the Upper Bound Theorem at work, sorting through the possible modes to find the one of the particular family which is the most likely to occur.

Consider the lower bound theorem however. To apply this, a detailed description of the moments at all points of the plate must be developed, and this description must then be checked to ensure that the yield condition is not exceeded (violated) anywhere, and also that equilibrium is satisfied.

Earlier, in Chapter 3, the equilibrium equations for the circular plate were established. These will now be used to develop a suitable description of the bending moments. From (3.3.3) the M_{rr}, $M_{\theta\theta}$ principal moments and transverse shear Q_r are related by

$$r(M_{rr})_{,r} - M_{\theta\theta} - rQ_r = 0, \qquad (4.4.4)$$

and

$$(rQ_r)_{,r} = pr.$$

The transverse share force, Q_r, can be eliminated to give

$$r(M_{rr}),_{rr} - M_{\theta\theta},_r = p_r. \quad (4.4.5)$$

The approach to using the lower the lower bound theorem successfully is usually made easy if some features of the *upper* bound solution, which would be sought first, can be incorporated. Here is it noted that the conical shape implied by the least upper bound is associated with a conical deflected shape and this in turn implies that there is yielding across radial lines, in order that the κ_2 curvatures can develop in the cone.

Hence it is concluded that it is likely that $M_{\theta\theta}$ will be at yield, i.e. $M_{\theta\theta} = -M_p$ at all points along a radius. When used in (4.4.5), this information gives a single equation for M_{rr} from which integration there is obtained

$$rM_{rr} = \frac{pr^3}{6} + Ar + B. \quad (4.4.6)$$

The $M_{\theta\theta}$ term, note, is lost in (4.4.5) because it is a constant. Now to find the values for A and B in (4.4.6)—at r = O, $M_{rr} = M_{\theta\theta}$ from symmetry, and hence M_{rr} (O) = $-M_p$.

Thus

$$r = O, \quad O = O + O + B,$$

but then $M_{rr}(O) = O + A = -M_p$.

Hence

$$M_{rr} = \frac{pr^2}{6} - M_p. \quad (4.4.7)$$

At the outer edge the radial moment is a maximum, but now hogging rather than sagging, and hence

$$M_{rr}(R) = +M_p.$$

Finally then

$$\frac{pR^2}{6} = 2M_p \quad \text{or} \quad P_L = \frac{12M_p}{R^2}. \quad (4.4.8)$$

This p_L is a *lower* bound estimate for the collapse pressure. It will be noted that it coincides in value with the least upper bound which has been constructed. The conclusion to be drawn is that since the upper and lower bounds coincide, so the value $p_c = 12M_p/R^2$ must be the *exact collapse load.* This example has been considered at length because of the importance of the solution being *exact.* In very few other instances will this prove possible.

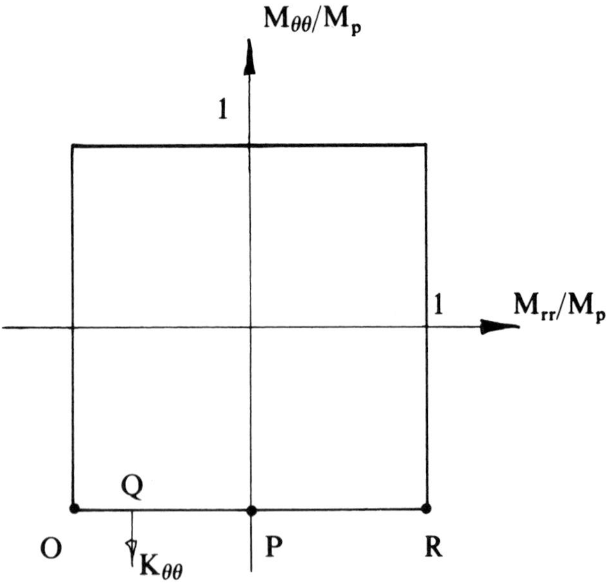

Figure 4.8 Square yield criterion—polar coordinates.

Before leaving this example, consider the role of the Normality Rule. In the evaluation of the moments for the lower bound solution, the convention has been adopted that hogging moments are positive. Suppose we ask whether the moment combination of this analysis can be traced out on the yield locus to any purpose?

Consider the vertex O (Fig. 4.8). The coordinates of this point are those for the moments at the centre of the plate. Successive points along the lower edge of the Yield Locus are the coordinates of the moment components for successive points along a typical radius, with the vertex R being the combination of moments which occurs at the edge of the plate. Note, however, that according to (4.4.7) the M_{rr} value varies parabolically and so the point $M_{rr} = 0$, $M_{\theta\theta} = -M_p$, namely point P, Fig. 4.8, corresponds to a point on the actual plate at radius r such that

$$0 = +\frac{p_c r^2}{6} - M_p, \quad \text{with} \quad p_c = \frac{12M_p}{R^2}.$$

Hence

$$r^2 = \frac{R^2}{2}$$

or

$$r = 0.707R. \tag{4.4.9}$$

What can the normality rule contribute? Draw the normal to the yield locus for some typical point, such as Q (Fig. 4.8). Note that this normal points wholly in the direction of $-M_{\theta\theta}$, and, because of the shape of the yield locus, has no component in the M_{rr} direction. The interpretation is therefore that the *allowable curvatures* according to the normality rule must be such that $\kappa_{\theta\theta}$ exists but κ_{rr} is zero. The $\kappa_{\theta\theta}$ exists but the *value* is not determined. This is consistent with the appearance of an arbitrary Δ, central displacement, in the upper bound solution. In effect any value of Δ and hence of relative $\kappa_{\theta\theta}$ values can occur, which corresponds to plasticity having set in and progressed to just such an extent as to have produced Δ.

Now, referring to the earlier discussion of surface curvature, the requirement that $\kappa_{\theta\theta}$ be present but κ_{rr} be zero can be seen to be consistent with the *conical* shape which has been produced as most likely by the upper bound calculations. It is therefore seen that the link between the two bounds has now been established quite firmly, since the lower bound use of the yield locus has suggested the use of the normality rule and this in turn has suggested the conical deformation mode, which was quite independently produced by systematic application of the upper bound theorem.

Exercise: Consider a simply supported circular plate, radius R, pressure loaded. Show that the exact solution for p_c is given by

$$\frac{p_c R^2}{M_p} = 6.$$

Confirm that the mode of collapse is conical and that the moments are everywhere sagging.

4.5 Circular plates–Tresca yield locus

In this section the problem considered in 4.4 will be reworked, the only change being the adoption of the Tresca rather than the square yield criterion (see Fig. 4.9). Differences will be apparent as the solution is developed, and clearly these will arise from the associated yet different yield criterion being used. First note that the square criterion circumscribes the Tresca locus. From this observation it can be concluded that the upper bound calculation in 4.4 can be carried directly over to this problem and will still represent an upper bound, but probably a less good one. That is to say, since the yield locus essentially implied by the upper bound calculation of 4.4 assumes $M = M_p$ along all yield lines, this may be an overestimate of the available yield capacity and hence may suggest a higher carrying capacity for the plate than is available when the material responds according to the Tresca Locus. This will be particularly true in regions of the plate where the principal (M_{rr}, $M_{\theta\theta}$) moments are of opposite sign, namely near the outer edge.

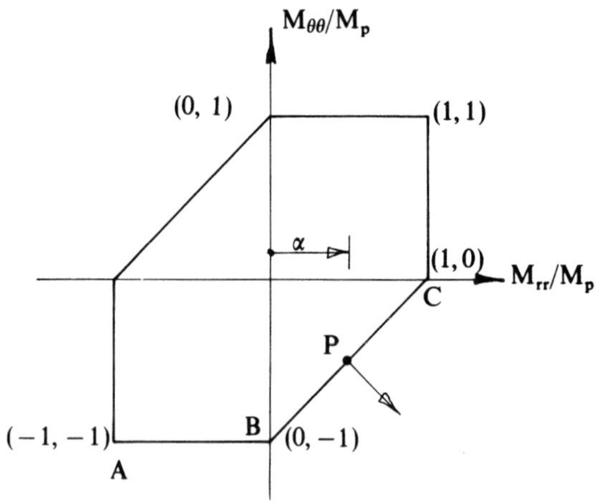

Figure 4.9 Tresca yield criterion–polar coordinates.

What then can be said about the lower bound calculation? The solution can begin much as before by noting that at the centre of the plate the moment field will be sagging and at the centre the moments can be thought of as described by vertex A, Fig. 4.9. Indeed, the plate inside some radius $\rho < R$, to be determined such that $M_{rr}(\rho) = 0$, behaves exactly as does a similar plate responding according to the square yield locus. It is expected that a conical mode shape will develop for $0 < r < \rho$, and the moments will be given by

$$M_{\theta\theta} = -M_p, \quad M_{rr} = \frac{pr^2}{6} - M_p. \tag{4.5.0}$$

Hence ρ is given by $M_{rr}(\rho) = 0$ or

$$\rho^2 = \frac{6M_p}{p}. \tag{4.5.1}$$

Recall that p is presently an unknown: indeed the primary object is to evaluate p, either as p_L or p_c, the lower bound pressure or the true collapse pressure, should it prove possible to satisfy all requirements of the problem. Thus ρ will not be 0.707R as in the previous problem where $p = 12M_p/R^2$; p is now expected to be *less* than this value because the yield locus can be *inscribed* in the previous yield locus.

For radii $\rho \leqslant r \leqslant R$ a new solution must be sought to the equilibrium equations which incorporates accurately the requirements of the Tresca Yield Locus. Now

$r = \rho$ corresponds to vertex B on the Yield Locus. As r extends beyond ρ it is tentatively proposed, to be checked later, that the moment combinations correspond to successive points along the segment BC of the yield locus. If this is so then it must be the case that for

$$\rho \leq r \leq R, \quad M_{rr} - M_{\theta\theta} = M_p,$$

since this is the equation of the face BC of the yield locus. But equilibrium still demands that $rQ_r = pr^2/2 + A$ and $A = 0$ since there are no point loads present.
Thus

$$(rM_{rr})_{,r} - M_{\theta\theta} = \frac{pr^2}{2}$$

and hence

$$(rM_{rr})_{,r} - (M_{rr} - M_p) = \frac{pr^2}{2}.$$

There is a cancellation of M_{rr} terms and on cancelling by r then integrating, we obtain

$$M_{rr} = \frac{pr^2}{4} - M_p \ln r + \alpha. \tag{4.5.2}$$

The region interior to $r = \rho$ must match with that exterior to $r = \rho$, at least in respect of the values of M_{rr} and Q_r. Hence

$$M_{rr}(\rho) = \frac{p\rho^2}{4} - M_p \,.\, \ln r + \alpha = 0 \quad (M_{\theta\theta} = -M_p) \tag{4.5.3}$$

and

$$Q_r(\rho) = \frac{p\rho}{2}$$

from either side of $r = \rho$.

The first of (4.5.3) is an equation for α in terms of p, ρ and M_p. Thus

$$-\alpha = \frac{p\rho^2}{4} - M_p \ln \rho = 1.5M_p - M_p \ln \rho \tag{4.5.4}$$

after (4.5.1) has been used.

But at $r = R$, the outer edge, suppose the M_{rr} value has reached some value associated with a point such as P (Fig. 4.10) with coordinates $(\alpha, -1 + \alpha)$.

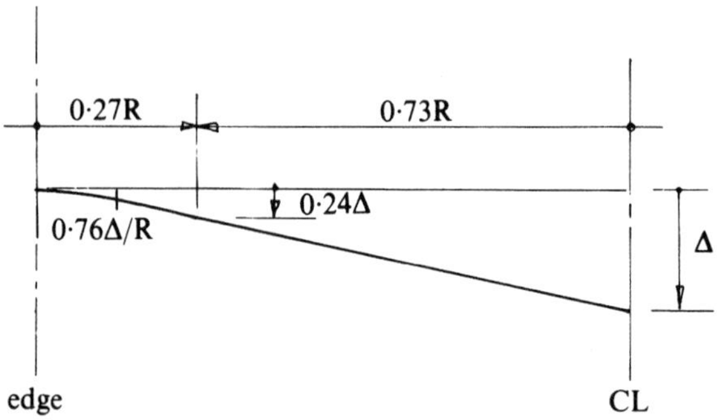

Figure 4.10 Tresca plate—section at collapse.

Then

$$\alpha M_p = \frac{pR^2}{4} - M_p . \ln R - (1.5 - \ln \rho)\, M_p,$$

$$= M_p \left[1.5 \left(\frac{R}{\rho} \right)^2 - \ln \frac{R}{\rho} - 1.5 \right]. \qquad (4.5.5)$$

In the spirit of a lower bound solution, the aim is to maximize p, i.e., minimize ρ. Expressed alternatively, (4.5.5) is an equation for R/ρ and the aim is to maximize this value. This is achieved when α is at its maximum of 1.
Then

$$1.5 \left(\frac{R}{\rho} \right)^2 - \ln \frac{R}{\rho} = 2.5$$

or $R/\rho = 1.37$, $\rho/R = 0.73$, and hence, from (4.5.1)

$$p_L = \frac{6M_p}{\rho^2} = \frac{6M_p (1.37)^2}{R^2} = 11.25 \frac{M_p}{R^2}. \qquad (4.5.6)$$

As observed earlier, this lower bound value is less than that associated with the square yield locus example for the reason that the material is weaker if responding according to the Tresca-type yield behaviour, for the same value of M_p.

Of perhaps more interest than the p_L value is the use of the normality rule to investigate the *shape* into which the plate is deformed in this case.

First note that there is a central conical region described by the normal to the yield locus having no component in the κ_{rr} direction and hence $\kappa_{rr} = 0$.

Now from earlier this is seen to require that

$$w,_{rr} = 0, \quad \text{or} \quad w = \alpha r + \beta. \tag{4.5.7}$$

In the outer region, for $r \geqslant \rho = 0.73R$, the normal to the yield surface between B and C is now seen to be equally inclined to the principal moment axes. More precisely, this normal, taking account of signs, is such that $\kappa_{\theta\theta}$ and $-\kappa_{rr}$ changes must be equal.

Hence

$$\kappa_{\theta\theta} = -\kappa_{rr}.$$

Expressed in terms of the transverse displacement and when the shallowness and surface of revolution properties are included, then

$$\frac{1}{r} w,_r = -w,_{rr}$$

or

$$w,_{rr} + \frac{1}{r} w,_{rr} = 0. \tag{4.5.8}$$

This is a differential equation for w which can be solved as

$$(rw,_r),_r = 0 \quad \text{or} \quad rw,_r = A$$

Hence

$$w = A \ln r + B. \tag{4.5.9}$$

The constants A and B can be found, in terms of the overall central displacement of the plate, Δ, by matching the displacement and slope of the plate at $r = \rho$ and noting that $w(R) = 0$.

Thus from (4.5.7, 8)

$$\beta = \Delta, \quad A \ln R + B = 0,$$

$$A \ln \rho + B = \alpha\rho + \Delta$$

and

$$\frac{A}{\rho} = \alpha\ .\ (\rho = 0.73R).$$

From these four equations

$$A \ln 1.37 = -A - \Delta \quad \text{or} \quad (1 + 0.315)A = -\Delta.$$

Hence, $A = -0.76\Delta$, $\alpha = -1.045/R$, $B = A \ln R = 0.760\Delta\ .\ \ln R$.

Displacement at $r = \rho$

$$\begin{aligned} w(\rho) &= -0.760\Delta \ln \rho + 0.760\Delta \ln R \\ &= 0.760\Delta \ln \frac{R}{\rho} \\ &= 0.24\Delta \end{aligned}$$

and hence the section on the deflected plate middle surface is as shown in Fig. 4.10.

To summarize, the value of p_L (4.5.6) has now been shown to be p_c, the collapse load, since a mode of deformation has been associated with the lower bound solution. If the solutions derived in (4.4) and (4.5) are compared, the effect of a change in yield criterion can be seen to lower the collapse load because the Tresca Yield criterion can be inscribed inside the other (square) criterion. The deformed shape changes in the outer zone from conical (square locus) to a "hat-brim" sort of shape (Tresca), the result of the inclined face BC on the yield criterion, Fig.4.9.

These two sections (4.4, 4.5) are important because they describe two of the few known exact solutions to plastic plate problems. In another sense they are important too—the circular plate represents a yardstick because the loaded area is enclosed by a minimum length boundary. Perhaps of more importance, the circular plate under pressure loading, and (say) a simply supported edge, is the *weakest* plate of given plan area (A) and given section strength (M_p). Expressed another way, the quantity $p_c A/M_p$ takes on a *minimum* value for the circular plate. It is not intended to prove this result, but intuitive arguments can be put forward to support the observation. The consequence of any such discussion is to further focus attention on the importance of the circular plate solution.

Exercise: Confirm that the solution for the simply supported circular plate, pressure loaded and subject to the Tresca yield criterion is identical to that for the square yield criterion.

4.6 Plates of other shapes—square and regular shapes

The circular plate has been discussed in some detail, because this problem will form a standard against which others can be compared. In this section other shapes will be considered, though in less detail, and mostly only as upper bounds on collapse load. The shape of most interest here is the rectangle.

Throughout this section it will be assumed that the plates are made from material which is isotropic and yields according to the *square* yield locus in principal moment space. This choice ensures minimum analytic complication while retaining all the essential features of collapse load calculation.

The bound theorems will be used, whenever appropriate, though the upper bound calculation will be seen to be much easier to perform. In a minority of practical examples only will it be found feasible to construct useful lower bound solutions for the collapse pressure. Consider then the problem of a quare, uniform, simply supported plate, side length L, loaded with a uniform pressure (p). The object is to calculate the collapse value of p.

Begin with an upper bound calculation. The simplest mechanism is a system of diagonal yield lines radiating into the plate corners. Suppose the central deflection is an arbitrary amount, Δ. If the plate is deforming plastically, then the calculation about to be made is an equilibrium one, which balances the work absorbed in the assumed hinge lines by the internal moments with he load work given up as the central deflection *increases* (from zero) by some arbitrary amount, Δ.

Thus the (small) angle α developed across a yield line is $\alpha = 2\sqrt{2}\Delta/L$ and the moment work will be the product of the yield moment (m)/unit length, times the length of line, times the rotation angle developed. Note that the angle α is constant along each line. This amount of work absorbed is supplied by the decrease in potential energy of the load (p), and this is the area integral of p $\times$ local transverse displacement. This reduces to p $\times$ volume swept out by the plate, as it deflects from the originally plane shape to the final pyramid shape.

Hence

$$4 \,.\, m \,.\, \frac{2\sqrt{2}\Delta}{L} \,.\, \frac{L}{\sqrt{2}} = \frac{1}{3} \,.\, p.L^2\Delta$$

or

$$\frac{p_1 L^2}{m} = 24. \qquad (4.6.0)$$

This is upper bound estimate of p_c, the collapse pressure. Hence

$$p_1 \geqslant p_c.$$

In order to gauge by how much p_1 is in excess of the exact p_c it is necessary to assume some more comprehensive mechanism which contains some parameters describing the deflected shape, then choose values for the parameters which will give a minimum value for p associated with this mechanism. These procedures will be illustrated in later sections. Here it is proposed to attempt the construction of an equilibrium moment field which can be used in constructing a lower bound estimate of p_c.

Choose an origin at the plate centre with axes x, y parallel to the plate edges. Then consider the bending moment distributions

$$M_{xx} = -\frac{pL^2}{24} + \frac{px^2}{6}$$

$$M_{xy} = +\frac{pxy}{6} \tag{4.6.1}$$

$$M_{yy} = -\frac{pL^2}{24} + \frac{py^2}{6}.$$

It is not obvious where these expressions have been derived from, but consider the form of the expressions. First along the edges $x = \pm L/2$, $M_{xx} = 0$ as required by the simple support boundary conditions. Indeed all the moment boundary conditions are met. Further, these expressions (4.6.1) can be shown, by direct substitution, to satisfy the plate equation (2.4.3). The loading at any point is carried equally by the bending in two directions and the twisting moments.

Consider further the values of the principal moments and where the largest principal values might be. For example, at the origin $M_{xx} = M_{yy} = -pL^2/24$, $M_{xy} = 0$. At a $M_{xx} = M_{yy} = 0$, $M_{xy} = pL^2/24$. If this system of moments is examined further it will be found that along any radial line the larger principal moment occurs for bending *about* the radial line and has a value equal to $pL^2/24$. Hence if

$$\frac{p_L L^2}{24} = M_p \tag{4.6.2}$$

then the yield condition is not violated anywhere. Indeed it is *met* in at least one direction *everywhere*. This p_L is a lower bound and will be seen to be equal to the upper bound value p_u (4.6.0). The conclusion to be drawn is that the *exact* solution p_c is also given by the same value

$$\frac{p_c L^2}{M_p} = 24. \tag{4.6.3}$$

Hence the square, simply supported pressure loaded plate of side L will collapse under this value of pressure, if the *square* yield locus applies. Indeed it can be shown that any *regular* n-sided ($n \geqslant 3$) *simply supported plate* has an exact solution for p_c given by

$$\frac{p_c R^2}{M_p} = 6 \sec^2 \frac{\pi}{n} \tag{4.6.4}$$

where R is the *radius* of the circumscribing circle.

Exercises: (1) The exploration of formula (4.6.4) is left as an exercise for the reader. Note that (4.6.4) includes (4.6.3) and the exercise in 4.4 as special cases.

(2) Repeat the upper bound calculations for collapse load on the square clamped plate *under pressure loading* and show that

$$\frac{p_u L^2}{M_p} = 48.$$

Now no simple lower bound solution is available (but see Chapter 6).

(3) The circular plate, either simply supported or clamped, and pressure loaded, represents a very important standard against which other shapes can be compared. Confine attention to plates which obey the square yield criterion. Then explore the proposition that the parameter $\gamma = p_c A/M_p$ takes on a *minimum* value for the circular plate. Here p_c is the collapse pressure, A is the plan area of the plate, or portion of it, which deforms, and M_p is the full plastic moment/unit length of the isotropic plate.

If the plate is simply supported then the minimum value is 6π, if clamped the minimum value is 12π. Typically, the solution for the square, simply supported plate has already been shown to produce a γ value of 24, which is greater then 6π.

Especially, consider some shapes which lack total symmetry.

II REINFORCED CONCRETE SLABS—UPPER BOUNDS

4.7 Yield line theory—I. Fundamentals—mainly isotropic

Thus far the theory and examples presented for plastic plates have been concerned with the homogeneous type most typically represented by solid metal plates. In this section the interest is in reinforced concrete plates of the type often encountered for example in floor systems in buildings.

Though there are many points of similarity in method between metal and concrete plates, there are also important differences, which are the justification for a separate treatment. The most important of these is in the material response to bending moment.

The physical situation is that the resistance of a concrete plate to bending resides in the reinforcement provided. Reinforcement parallel to the x axis provides resistance principally in the M_{xx} direction, that is about an axis normal to the x direction. Resolved on to an axis which makes an angle of θ with the M_{xx} axis, this moment of resistance provides a moment of $M_{xx} \cdot \cos^2\theta$. Thus the moment of resistance resolves as a second order (tensor) quantity and not as a vector (first order tensor). Recall that vector quantities resolve as $\cos\theta$ and not $\cos^2\theta$.

If more than one system of reinforcing bars is incorporated, then the resistance moment across any chosen direction is taken to be the summation of the resistance

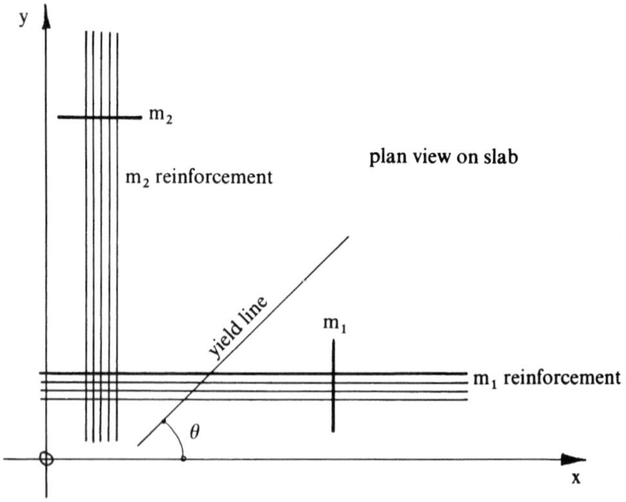

Figure 4.11 Reinforcement and moment capacities.

moments provided in this direction according to the expression just quoted. If the direction is not principal then twisting moments will also act. In general twisting moments however play little part in the mechanics of concrete plates at ultimate load.

Consider an orthogonal system of reinforcement with moment of resistance m_1 in the x direction and m_2 in the y direction, for sagging bending moment. Then the *axis* of the m_1 moment of resistance will be parallel to the y axis; that is, normal to the reinforcement direction (Fig. 4.11). Likewise, for m_2 the axis for bending is parallel to the x axis. Suppose for the present that $m_1 \neq m_2$; then the system is described as *orthotropic*.

For a line making an angle θ with the x axis (Fig. 4.11) the combined moment strength *about* this line of the two systems of reinforcement is given by

$$\begin{aligned} m_\theta &= m_2 \cos^2 \theta + m_1 \cos^2\left(\frac{\pi}{2} - \theta\right), \\ &= m_2 \cos^2 \theta + m_1 \sin^2 \theta. \end{aligned} \tag{4.7.0}$$

In the special case when $m_1 = m_2$, then this orthogonal arrangement of reinforcement is called *isotropic* and (4.7.0) shows that

$$m_\theta = m_1 \cos^2\theta + m_1 \sin^2\theta = m_1 \equiv M_p, \tag{4.7.1}$$

namely the strength of the slab is equal in all directions. Isotropic and orthotropic systems of reinforcement are the two most important systems encountered in practical applications.

The isotropic system described above can conveniently be represented in yield condition terms by the square yield criterion of earlier (4.4, Fig. 4.1) on axes of principal moment—the m_1 or m_2 (= m_1) axes. This yield criterion states that each of m_1 and m_2 can rise to M_p, the full plastic value, independently, and the yield condition is just satisfied. In the conventional study of concrete slabs by the yield line method, explicit reference to the yield locus (or criterion) is seldom made because the calculation made is an upper bound on the collapse load, and in this attention is focused on the mechanism and equilibrium conditions. The yield condition is *assumed* to be met across all of the assumed yield lines. This is not rigorously ensured, however, so the calculation can only be expected to be an upper bound on the collapse load.

The concept of a *mechanism of failure* has already been encountered (4.3, 4.4). This concept will now be extended and used in conjunction with a work calculation as the main solution process for slab problems by the *yield line method.*

A mechanism is a combination of directions in the slab about each of which rotation is assumed to occur, so producing movement from the original plane slab to some shallow pyramid-shaped surface. All the directions which will be considered here will be straight. This is adopted for simplicity and means that all the deflected surfaces will have plane faces which meet other plane faces along the yield lines. To be a possible mechanism there are certain requirements of the yield line arrangement, but it is not easy to set these down in the form of adequate rules.

Consider the corner of a slab (Fig. 4.12). If the slab in the corner is to deform by yield line rotation then rotation must also occur about the simply supported edges. Indeed these edges can be thought of as yield lines of zero strength. Hence the sagging yield line AB must radiate *into* the corner A, and rotation of leaf I of slab with respect of leaf II about AB is accompanied by rotation about AC and AD, allowing B to descend and relative rotation about AB to occur.

But it is possible for other arrangements of yield lines into the corner to occur, for example, the hogging line ab associated with the sagging lines ac, bc, cd. This is a valid system provided that dc (produced) passes through A, for unless this is so the rigid leaves of slab spanning on to the supports at aC and bD could not rotate. In this system of yield lines the hogging line (ab) has in effect become part of the boundary, and would be a line of zero moment strength if no hogging strength was provided.

Specifying the transverse displacement of d fixes the amounts of rotation of the slab portions about aC and bD and hence too the displacement of c. Then the rotation of abc about ab is also fixed. Hence all the movements and rotations relate back to that of d. This is spoken of as a *one degree of freedom* system. These are the most usual types of mechanism, though multi-degree of freedom systems are possible.

Yield line patterns may be specified by *parameters* and these parameters should be chosen to give as low an estimate of p as possible. For example a, b and c can be varied in position. But as a piece of practical advice it is suggested that as *few parameters as possible* be associated with a given mechanism—at least initially.

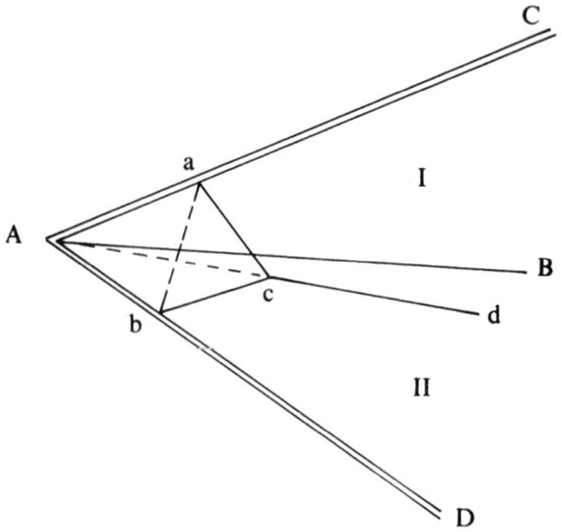

Figure 4.12 Yield lines—geometrical requirements.

Consider the class of rectangular simply supported slab shapes (Fig. 4.13*a*).

Then the crudest mechanism possible consists of two intersecting yield lines which run from one corner to an opposite corner. If it is assumed that the slab has no hogging resistance, then there is no restraint to the edges of the slab rotating about their supports. Suppose sagging yield occurs along these two diagonal yield lines to give a maximum transverse deflection at the centre point of Δ (say). Here Δ is the one degree of freedom. Then by rotation about the edge lines of the four respective portions of slab deflect into the shallow pyramid shape with maximum central deflection of Δ, (Fig. 4.13*a*).

In reality, if an experiment was to be conducted this might not be the actual collapse mechanism to develop. However, it is likely that an Upper Bound calculation based on this mechanism will be a reasonable estimate of the collapse load capacity of this slab, but an overestimate. From the geometry of deformation, with a central deflection of Δ, the (greatest) slopes of the respective portions of slab are $2\Delta/l$ and $2\Delta/\mathrm{L}$. These slopes are vector quantities. Resolving them normal to the supposed diagonal yield lines, there is obtained a common rotation about these yield lines of

$$= \frac{2\Delta}{l} \times \cos\alpha + \frac{2\Delta}{\mathrm{L}} \times \sin\alpha,$$

$$= \frac{2\Delta}{\sqrt{\mathrm{L}^2 + l^2}} \left[\frac{\mathrm{L}}{l} + \frac{l}{\mathrm{L}} \right]$$

Suppose the slab to be subjected to a uniform pressure loading, p_1. Suppose also that the slab is isotropically reinforced for sagging moments, to a strength

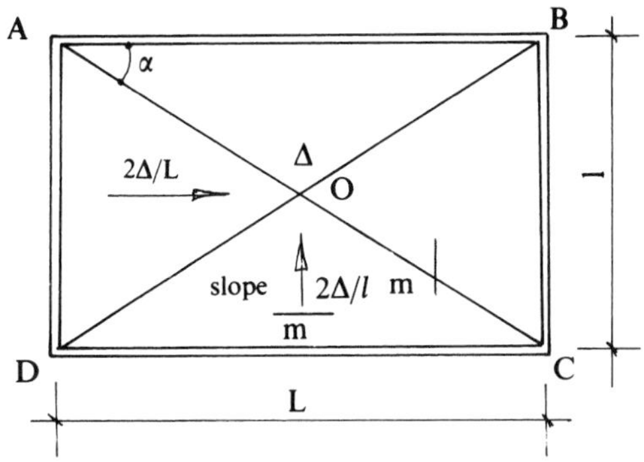

Figure 4.13(*a*) Example—rectangular plate I.

$m_1 = m_2 = M_p$ parallel to the slab edges. Then, by the analysis earlier in this section, the strength across any possible sagging yield line will also be M_p. Use a work calculation in which a virtual central displacement of Δ associated with the above mechanism produces internal work absorption along the yield lines; this is to be balanced by loss of potential energy of the loading. Then the virtual work balance will appear as

$$\frac{2\Delta}{\sqrt{L^2+l^2}}\left[\frac{L}{l}+\frac{l}{L}\right].\ M_p \times 2\sqrt{L^2+l^2} = \text{internal moment work absorbed}$$

$$= \text{load work done} = \frac{p}{3}\Delta L l$$

$$= (p \times \text{volume of pyramid shape}).$$

Hence, solving for p_1, and noting that the virtual Δ cancels, obtain

$$p_1 = 12\left[\frac{L}{l}+\frac{l}{L}\right]\frac{M_p}{Ll}$$

$$= 12\left[\frac{L^2+l^2}{(Ll)^2}\right]M_p. \qquad (4.7.2)$$

Below it is shown that this estimate for the true p_c is greater than that associated with another general layout of yield lines. Hence, it can be concluded that the second calculation (p_2) is superior to p_1 as an estimate for p_c, since an upper bound calculation for p_c is being performed, and the lowest current value of p is the *best* so far available.

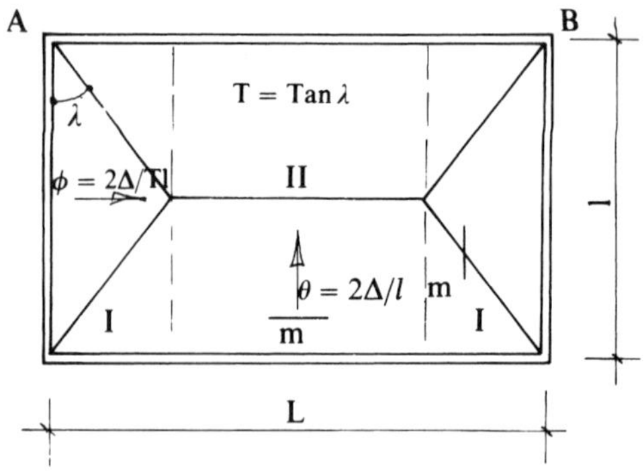

Figure 4.13(*b*) Example—rectangular plate II.

This second calculation is based on the mechanism shown in Fig. 4.13*b*. Here the symmetry about each of the mid-lines is preserved, but a more comprehensive mechanism chosen. This is still a mechanism with just *one* degree of freedom, thought of as Δ, but λ is an unspecified angle, which in the previous mechanism was fixed as $T = \tan\lambda = L/l$.

Leaving λ as a parameter, the external load work and moment work can be evaluated as previously to give:
External load work

$$= \frac{1}{3}(l \,.\, lT)p_2\Delta + (L - lT)\frac{1}{2} \,.\, p_2\Delta$$

$$= M_p\left[2l \,.\, \frac{2\Delta}{Tl} + 2L \,.\, \frac{2\Delta}{l}\right] = \text{ internal moment work.}$$

Here the evaluation has been arranged to split the load work so that the two zones I contribute equally to the first load work term, where zone II accounts for the second. The internal moment work can alternatively be derived exactly as in the previous calculation but as an angle of rotation of Φ_1 across the diagonal lines of

$$\Phi_1 = \frac{2\Delta}{l} \,.\, \sin\lambda + \frac{2\Delta}{Tl} \,.\, \cos\lambda$$

acting on moment M_p along a yield line of length 4 × AO, together with $\Phi_2 = 4\Delta/L$ acting on M_p along the central yield line of length $(L - lT)$.

Simplifying (4.7.3) and solving for p_2 there is obtained an expression still containing the parameter $T = \tan\lambda$. As λ is varied so p_2 will change. What is sought is the value of λ for which p_2 takes on a minimum value. Now

$$4M_p\left[\frac{1}{T}+\frac{L}{l}\right]=\frac{p_2 l^2}{6}\left[\frac{3L}{l}-T\right]. \tag{4.7.3}$$

This minimum for p_2 can most easily be sought by differentiating p_2, regarded as a function of **T**, with respect of T, rather than λ. There is then obtained a lengthy expression, since p_2 is a quotient in T. The minimum is associated with a value of zero for the *numerator* in this expression.

Hence

$$\left[\frac{1}{T}+\frac{L}{l}\right](-1)=\left[\frac{3L}{l}-T\right]\left(-\frac{1}{T^2}\right).$$

Simplifying, we obtain

$$\left(\frac{L}{l}\right)T^2+2T-\left(\frac{3L}{l}\right)=0.$$

This quadratic in T provides only one physically interesting root, namely

$$T=\frac{\sqrt{3(L/l)^2+1}-1}{(L/l)}. \tag{4.7.4}$$

Then, typically, for the case $L/l = 1$, $T = 1$ and

$$p_2 = 24\frac{m}{l^2}.$$

This is the same value as given by p_1 in (4.7.2).

But this should be so since for $L/l = 1$, the two mechanisms are identical.

However for $L/l > 1$, $p_2 < p_1$ as shown by Table 4.0, indicating that p_2 is the superior value.

The p_2 values have been found as the result of a minimization, involving differentiation and solution of a quadratic equation—and this with just a single parameter in the mechanism choice. In many practical problems it is not either feasible or worthwhile, in terms of the final result, to contemplate such a procedure.

Instead some reasonable choice of value for the free parameter, or some few reasonable choices, followed by an *evaluation* of the p expression should suffice to give an adequate exploration of the chosen mechanism. In respect of the example discussed above, a reasonable choice for T might be unity together with say 0.75 and 1.5, to give three values for p_2 and indicate the effect on p_2 of changes in T. By plotting p_2/T, the minimum or near minimum value can then usually be detected.

Table 4.0 Collapse pressures for two mechanisms

L/*l*	k_1	k_2
	$p_2 = k_2 \frac{M_p}{l^2}$ (Fig. 4.13*a*)	$p_2 = k_2 \frac{M_p}{l^2}$ (Fig. 4.13*b*)
1	24	24
2	15	14.1
3	13.3	11.7
α	12	8

In the example just discussed an implicit use of the so-called *projected lengths* formula has been made. Consider a piece at yield line ab between two pieces of slab, and consider the edges about which these pieces of slab I, II are rotating (Fig. 4.14). For ab to be a possible yield then ab extended must pass through the point where the two edges meet. Suppose the angle α, β define the position of the yield line with respect to these edges (Fig. 4.14). Then if θ, ϕ are the maximum slopes of the slab portions I, II, the angle between I, II across ab is given by $\theta.\cos\alpha + \phi.\cos\beta$, and the moment work along this line ab is M_p (ab) $(\theta\cos\alpha + \phi\cos\beta)$.

Rearrange this expression to read $M_p(\theta L_1 + \phi L_2)$, then this is the *projected lengths* formula for the piece of yield line ab. Here L_1, L_2 are the projections of ab on the two edges respectively. The technique then is to project the yield line onto the edges and sum the products of projected length × maximum slope.

As an aid to deciding whether the current estimate of p_u is satisfactory, the following procedure is suggested. The procedure described below is applicable only to slabs which are *uniformly supported* all round their edges, either as simply supported, or as clamped edge slabs. Excluded therefore are slabs with mixed edge conditions, or with unsupported free edges. For the present, too, the slabs will be assumed to be isotropic.

What is proposed is to compare the current slab under consideration with an *ideal* slab, namely the circular slab. Now the collapse pressure for a given strength and layout of slab can be expressed in the form

$$\text{collapse pressure} = \text{coefficient} \times \frac{M_p}{(\text{plan dimension})^2}.$$

This follows from the dimensions of the relevant quantities: the pressure has dimensions force/length^2, the moment of resistance per unit length has dimensions of force and the plan dimension those of length. Hence dimensionally the

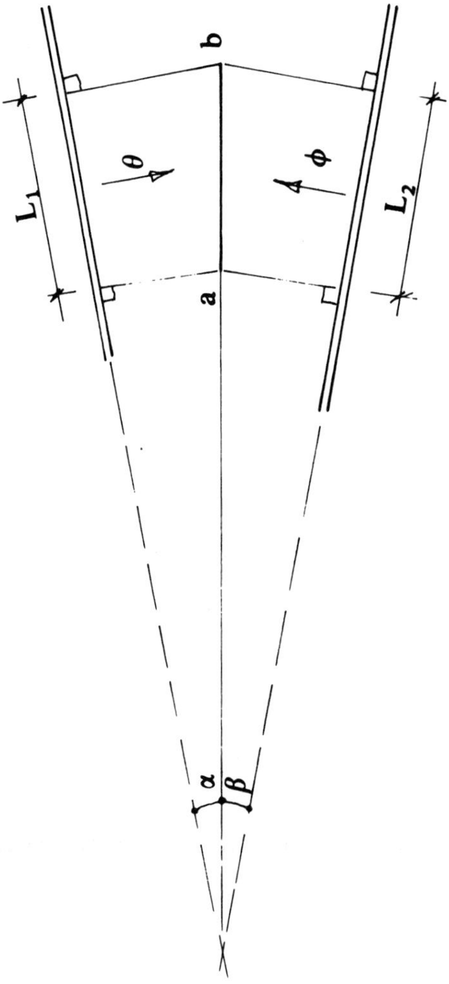

Figure 4.14 Projected lengths of a yield line.

above form is a permissible combination. However, it is proposed to make a specific choice to replace the (plan dimension)^2 term—the l^2 of the expressions derived earlier in this section—by the choice *slab plan area*, A, which also has dimensions L^2.

Hence the expression for the collapse pressure p_c will be sought in the form

$$p_c = \gamma \frac{M_p}{A} \tag{4.7.5}$$

where γ is a numerical coefficient, M_p is the isotropic moment of resistance of the slab and A is the *plan area* of the slab which is *actually spanned* by the chosen mechanism. Usually this will be the whole plan area of the slab.

For the simply supported circular plate the value of γ has already been shown (4.4) to be

$$\gamma = 6\pi = 18.85 \tag{4.7.6}$$

This coefficient γ is a *minimum* for the circular slab, though no proof of this will be offered here.

There is a useful parameter which describes the general *shape* of the slab. This is the parameter

$$\alpha = \frac{(\text{perimeter})^2}{\text{plan area}}. \tag{4.7.7}$$

Note that both α and γ are *pure numbers*. Now α also has a *minimum* value for the circular slab of $(2\pi/r)^2\ \pi r^2 = 4\pi = 12.57$. All other slab shapes have values of α in excess of 4π.

Consider finally a third non-dimensional parameter β given by

$$\beta \equiv \frac{\gamma}{\alpha}. \tag{4.7.8}$$

Then for the circular simply supported slab

$$B = \frac{6\pi}{4\pi} = 1.5. \tag{4.7.9}$$

It is an observation which can be tested with examples that β, evaluated for many simply supported isotropic plates under pressure loading, *is greater than or equal to 1.5*, and is usually in the range 1.5–1.6. Should it happen that the value of β is greater than about 1.6 then the slab is elongated in one direction and hence can probably be treated as almost a one-way spanning system.

This observation can be used in a variety of ways. First, either the collapse pressure (p_c) or the required slab strength (m) can be solved for if the other

Table 4.1 Rectangular slab —non-dimensional parameters (see Fig. 4.13*b*)

L/l	γ	α	β
1	24	16	1.50
2	28.3	18	1.57
3	35.2	21.3	1.65
4	42.7	25	1.71

quantities are known. Alternatively, computing the value of β at the conclusion of an independent calculation for p_c or M_p can serve as a quick and useful check on the correctness, or at least the plausibility, of the result.

For example, the rectangular slab discussed earlier in this section produces the β values in Table 4.1.

Exercise: The discussion thus far has dealt with simply supported slabs. Show that if the edges of the slab are restrained against rotation by hogging strength in the slab (M_p) of equal strength to the sagging strength (M_p), then the collapse pressure of an all fixed edge slab is *double* that of the simply supported equivalent.

In terms of the parameters γ, α, β, this means that γ will doubled and so too will β. Hence for fixed edge slabs, $\beta \geqslant 3.0$. For practical purposes an upper limit on β is say 3.2, beyond which the slab is sufficiently elongated to be treated as basically a one-way spanning system.

4.8 Yield line theory—II. Further isotropic examples

The slab will still be assumed to be isotropic in this section, but a sufficiently complex problem will be investigated to illustrate some of the computational arrangements which might be found useful in the tackling of a typical problem.

The simply supported slab shown in Fig. 4.15 is considered isotropic and loaded with a pressure loading. Assuming the slab strength to sagging moments to be M_p and hence this same value to apply along all yield lines by virtue of the isotropy, it is required to compute a sufficiently close upper bound to be confident that the exact collapse p_c is not being overestimated by more than a few per cent. The intention is to use a work calculation based on an assumed mechanism to calculate an upper bound collapse pressure p_1.

Consider first then what a suitable mechanism might be in this case. The slab is very roughly oval in shape; certainly it has a convex shape of edge line, seen

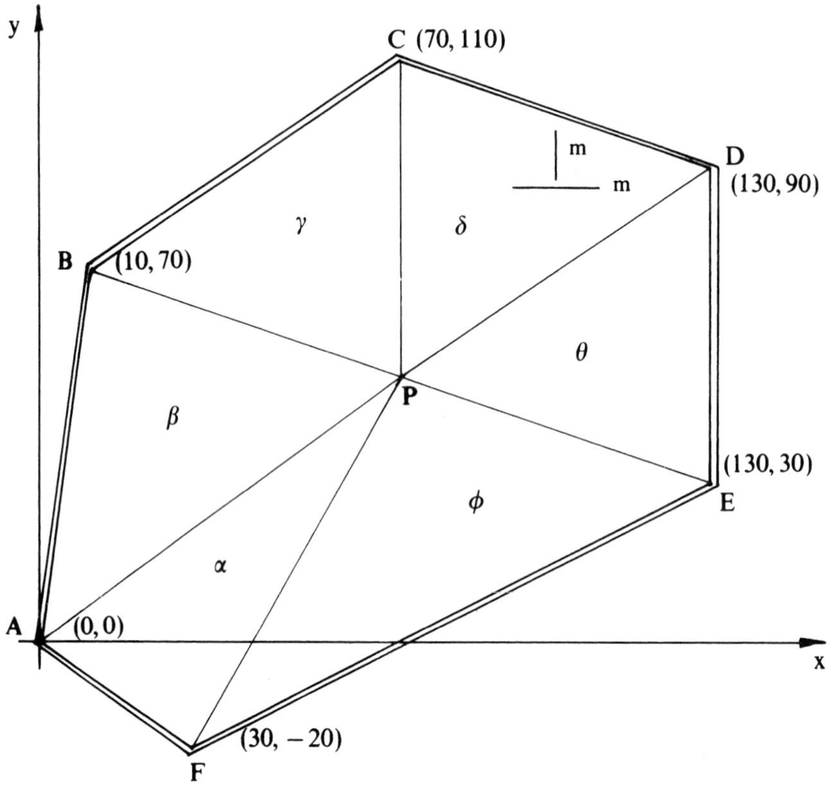

Figure 4.15 Irregular slab, simple mechanism.

from the outside. One possible mechanism is a pyramid shape based on some point (P) in the interior as the apex, with a yield line radiating to each corner. A useful piece of advice in such cases is to begin with the *simplest possible* mechanism in order to generate an answer of some kind. This is much more prudent and satisfying than approaching a more complicated mechanism and possibly even attempting a minimization, becoming bogged down in symbols and formulae and finding the enterprise impossible to complete. So let us begin with this simplest, though possibly rather crude approximation to the problem.

Where then to place the apex? For simplicity let this point (P) be on the join of vertices B and E. Let it be the point with coordinates (70, 50). The analysis of the problem will be written in terms of the actual dimensions. It is advised that the reader redraw Fig. 4.15 on graph paper and follow the steps by making check measurements at the various stages of the calculation.

The six segmental areas into which the slab is divided by the yield lines have been labelled $\alpha-\phi$. Measurements of the lengths of the supported edges give the values 36, 70.5, 72, 63, 60 and 112, for $\alpha-\phi$ respectively. Next the perpendicular distances of the apex (70, 50) from the respective edges are measured to be 80, 62, 50, 56.5, 60, 44. With these data the plan area of the slab can be evaluated as

$$\frac{1}{2}[\Sigma\, 36 \times 80 + \ldots] = 11{,}300 \equiv \text{A}.$$

The boundary perimeter, $B = \Sigma l_i = 413.5$.
Then the load work done $= 1/3\text{p}\,\Delta\text{A}$
and the moment work absorbed $= M_p\Sigma(l_i\theta_1)$

$$= M_p\left[\frac{36}{80}+\frac{70.5}{62}+\frac{72}{50}+\frac{63}{56.5}+\frac{60}{60}+\frac{112}{44}\right]\Delta$$

$$= 7.69 M_p \Delta.$$

$$\therefore \frac{\text{pA}}{M_p} = 3 \times 7.69 = 23.1\ .\ (=\gamma).$$

Now

$$\alpha \equiv \frac{B^2}{A} = 15.11 \quad \therefore \beta = \frac{\gamma}{\alpha} = \frac{23.1}{15.1} = 1.53.$$

Note the closeness of β to the lower limit of 1.50. This suggests that the point (70, 50) as apex is nearly the most advantageous choice. As a check, another vertex can be chosen, and the calculation quickly repeated. For example, if the point (50, 40) is chosen then the corresponding estimate for p is given by

$$\gamma = \frac{\text{pA}}{M_p} = 3 \times 7.89 = \underline{23.7}$$

and hence

$$\beta = \frac{\gamma}{\alpha} = \frac{23.7}{15.1} = 1.57.$$

This again confirms that the initial choice of point P (70, 50) was a sound one.

To complete the investigation of this example some other plausible mechanism should be chosen. This is left as an exercise.

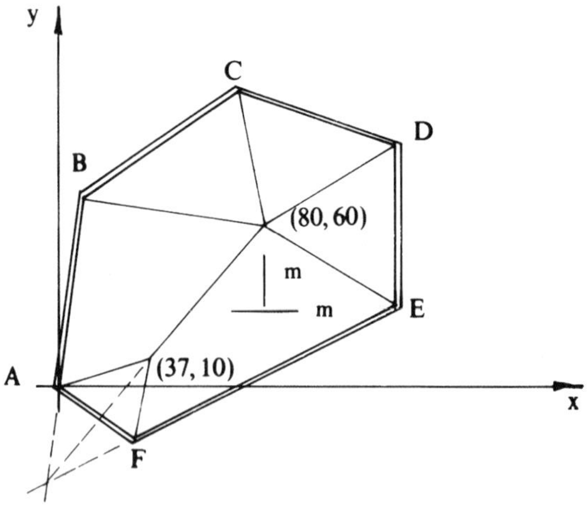

Figure 4.16 Example—second choice of mechanism.

Exercise: Repeat the above example but with the choice of mechanism as shown in Fig. 4.16. This is still a one degree of freedom mechanism, but the calculation is a little more complicated than the previous mechanisms associated with this shape. *Answer*: $\gamma = 23.4$; $\alpha = 15.5$; $\beta = 1.55$.

4.9 Yield line theory—III. Orthotropic problems

Thus far only isotropically reinforced slabs have been considered. If the reinforcement provided is *different* in the two orthogonal directions, then the system is no longer isotropic. Such slabs are referred to as *orthotropic slabs*. The yield line calculations are similar to those for isotropic slabs, but it must be remembered that now the moment strength along any particular yield line will depend upon how this line is oriented with respect to the now different slab strengths in the two orthogonal directions. Figure 4.11 (p. 102) shows a typical yield line crossing orthotropic reinforcement. The relevant yield moment about this line is given from (4.7.0) as

$$m_\theta = m_2 \,.\, \cos^2\theta + m_1 \sin^2\theta. \tag{4.9.0}$$

If $m_2 = \mu m_1$ then μ will be termed the *orthotropic ratio*. Generally μ will be in the range $0.25 \leqslant \mu \leqslant 4$ with the isotropic ratio providing a useful comparision value at $\mu = 1$.

Consider again the example (4.7) of the simply supported rectangular plate, loaded with a uniform pressure, but now reinforced orthotropically. Figure 4.17*a* is a redrawn version of Fig. 4.13*b*, with $L = 2l$ and $m_1 \neq m_2$. Calculation of the

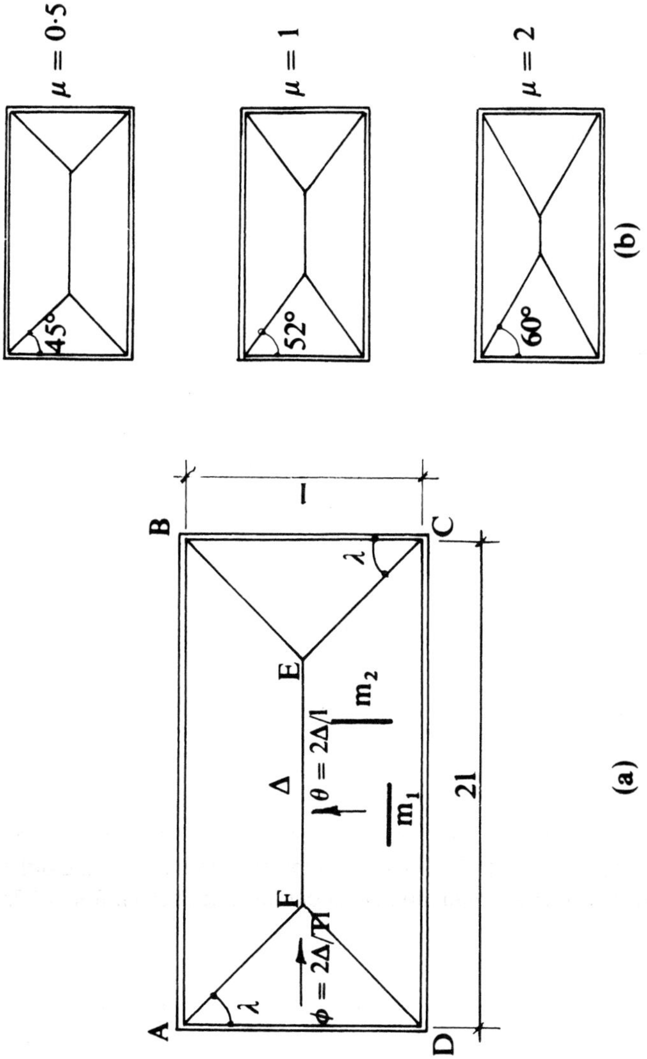

Figure 4.17 Orthotropic rectangular slab.

collapse pressure can proceed exactly as for the isotropic case, except that now the yield moment along the yield lines such as EC must be calculated from (4.9.0). In the present case, suppose $m_2 = m_1/2$, i.e. $\mu = 0.5$, then the yield moment along a typical line such as EC is given by

$$m_\lambda = m_1 \sin^2 \lambda + m_2 \cos^2 \lambda = m_1(\sin^2 \lambda + 0.5 \cos^2\lambda). \qquad (4.9.1)$$

Recall that λ is an unknown, to be found from the condition that the collapse pressure (p) associated with a particular λ should be a *minimum*. While it is in principle possible to proceed with an algebraic investigation of such problems, it is strongly recommended that instead, likely values of λ be chosen, the corresponding p values calculated and the *least* p then chosen as the relevant value. Such a method is less elegant but much more likely to systematically identify the minimum value.

It is useful to keep in mind an *associated isotropic* slab—namely that slab which contains the same amount of reinforcement as the orthotropic one. To a good approximation the *amount* of reinforcement is *proportional* to m, the moment yield strength. Also recall that m is expressed per unit width of slab.

Hence the total amount of reinforcement for an isotropic slab is proportional to

$$2M_p \times \text{slab area} \equiv V_M. \qquad (4.9.2)$$

This result follows because there are *two* directions of reinforcement and each direction requires an amount proportional to the slab area.

For the orthotropic slab

$$V_M = (m_1 + m_2) \times \text{slab area}. \qquad (4.9.3)$$

The symbol V here, and in Chapter 5, is used to denote the total amount of reinforcement.

Hence

$$m_1 + m_2 = 2m \qquad (4.9.4)$$

is the required equivalence relation. Now $m_2 = \mu m_1$.

Thus

$$m_1 = \frac{2m}{1+\mu},$$

$$m_2 = \frac{2\mu m}{1+\mu}, \qquad (4.9.5)$$

for an orthotropic slab to contain the same amount of reinforcement as the equivalent isotropic slab. In the present case $\mu = 0.5$ and hence

$$m_1 = 1.33m, \quad m_2 = 0.67m. \qquad (4.9.6)$$

For the mechanism shown in Fig. 4.17(*a*)

$$4m_1 (2 - T) + 4m_\lambda . (T + 1/T) = pl^2 (1 - T/6), \tag{4.9.7}$$

where T = tan λ. Then (4.9.7) is the orthotropic equivalent of (4.7.3).

What is recommended is that trial values of λ are used to evaluate p. For example, if λ = 45° then from (4.9.1), (4.9.6) m_1 = 1.33m, m_λ = 0.75 × 1.33m = m. But T = 1 and hence from (4.9.7) pl^2/m = 16. Other values of λ can be chosen and the evaluation of p repeated. Thus λ = 40° leads to m_λ = 0.94 m and pl^2/m = 16.1, while λ = 50° gives m_λ = 1.05m and pl^2/m = 16.1. There is evidently a very shallow minimum for p at about λ = 45°.

What then has been established? It has been shown that if the reinforcement in the rectangular slab, side ratio 2:1, is arranged with the moment strength *across* the shorter span to be *double* that along the slab length, namely μ = 0.5, but with the total amount of reinforcement constrained to be the *same* as that required for an isotropic slab of strength m, then

$$\left(\frac{pl^2}{m}\right)_{\mu=0.5} = 16, \left(m_1 = \frac{4}{3}m, m_2 = \frac{m_1}{2}\right)$$

and from Table 4.0,

$$\left(\frac{pl^2}{m}\right)_{\mu=1} = 14.1. \tag{4.9.8}$$

In other words, the orthotropic slab can sustain some 13% more load than the associated isotropic slab. There is a catch here, however. Suppose instead of μ = 0.5 the ratio μ = 2 had been used, then the same amount of reinforcement is implied but now

$$m_2 = 1.33m, \quad \text{and} \quad m_1 = 0.67m.$$

The roles of the two directions are exchanged and

$$M_\lambda = m_2 (\cos^2 \lambda + 0.5 \sin^2 \lambda). \tag{4.9.9}$$

For λ = 45° the value of m_λ is exactly the same for μ = 0.5, namely m_λ = m but the associated collapse pressure from (4.9.7) is less, pl^2/m = 12.8. The minimum value of p is now when λ = 60° with pl^2/m = 11.85.

Hence the comparison to be made is

$$\begin{aligned} \mu &= 0.5, & pl^2/m &= 16, \\ \mu &= 1, & pl^2/m &= 14.1, \\ \mu &= 2, & pl^2/m &= 11.85. \end{aligned} \tag{4.9.10}$$

The $\mu = \frac{1}{2}$, 2 cases are coupled since they imply equal total amounts of reinforcement, and this amount has been chosen to be the same as for $\mu = 1$, the associated isotropic case. Hence it is seen that there is a correct and an incorrect way to introduce orthotropy. In the present case it is easy to conclude on intuitive grounds that $\mu = 0.5$ is to be preferred to $\mu = 2$, but in more complicated examples this may not be so easy. It is suggested therefore that the associated isotropic value of $\mu = 1$ should always be computed and then used as a measure of the advantage to be gained by the proposed orthotropic layout. Should it happen that the orthotropic strength is *less* than the isotropic one, then change the ratio to be the inverse of the previous value and repeat the calculation. This will lead to a strength advantage for the new orthotropic arrangement.

In Fig. 4.17*b* are shown the positions of the critical yield lines for the three values of μ studied. As the short span is made stronger at the expense of the long span, so the system becomes more *one* dimensional and stronger, with the corner lines being confined to smaller areas of the slab. If even more extreme values of μ are chosen it can be shown that a maximum (theoretical) advantage is gained at around $\mu = 0.1$ of about 24% pressure increase for the orthotropic slab. However, in practical cases such a ratio would probably lead to unsightly cracking parallel to the short side in the end regions.

By adopting the system of trial choices for λ and evaluation of p, systems of almost any complication can be handled. The reinforcement need not be orthogonal, and different ratios may be adopted for the bottom and top steels. An alternative and more elegant approach would have been to introduce the *affinity* theorems of Johansen (see Section 4.14, below). There are limitations to use of the theorems however and some aspects of the physical situation are lost in the process, and so these theorems are not included here.

One final comment on orthotropy—the non-dimensional parameters α, β, γ introduced in (4.7) can be applied to orthotropic slabs, through the associated isotropic slab. Thus for the $\mu = 0.5$ case

$$\gamma = pA/m = 2 \times pl^2/m = 32,$$

and

$$\alpha = (6l)^2/2l^2 = 18.$$

Hence

$$\beta = \frac{\gamma}{a} = 1.78 > 1.5$$

On the other hand the $\mu = 2$ example gives

$$\gamma = 23.70, \quad \alpha = 18,$$

and hence

$$\beta = 1.32 < 1.5.$$

That $\beta < 1.5$ is another signal showing that $\mu = 2$ is a disadvantageous orthotropic ratio. Instead $\mu = \frac{1}{2}$ should be chosen, with no reinforcement penalty, but a strength advantage.

4.10 Strip method—Hillerborg's proposals

In recent years a further method has been gaining popularity in dealing with slab design questions. The initial proposals were made by A. Hillerborg in about 1960 and have been elaborated since. Here only a brief sketch is possible and all the discussion will be limited to rectangular slabs.

From a practical viewpoint, rectangular slabs are likely to be reinforced along directions parallel to the slab edges, even although these directions are unlikely to be principal moment directions. From what has been seen of slab problems so far, it will be evident that many comparatively simple and practical problems cannot be fully solved whether elastic or plastic. In the case of elastic plates, there are only a few shapes capable of analytical solutions. Most data have been accumulated through numerical solutions, notably finite element analyses in recent years. In the case of plastic plates, upper bounds on the collapse pressure are readily found but few associated lower bounds are known.

The Hillerborg proposals aim to resolve this latter stalemate by *assuming* that the (rectangular) slab does not support *any* twisting moment in the x, y directions. In effect the slab is notionally thought of as composed of a grid-work of beams which in some manner interact with one another to carry the imposed loading.

What is aimed at is an equilibrium system of bending moments which support the given load, and hence the method is in the spirit of a lower bound on the collapse pressure. In the simplest form the method assumes all the load is dispersed to the boundaries by beam strips in both x and y directions (Fig. 4.18). A decision is needed on how the load should be divided up. On Fig. 4.18 are shown some dotted lines and arrows. The dotted lines divide the whole area up into zones, and the *whole load* within each zone is then *assumed* to be carried by strips in the direction of the arrow. Some judgement is required in making this subdivision. For the rectangular slab illustrated, $\theta = \pi/4$ is a reasonable choice. This choice is backed up by prior knowledge of the corresponding elastic solution.

Now a typical x direction beam strip will have loading near its ends with none in the central section, whereas a typical y direction beam strip will carry a full load if near the slab centre, or end loading only for a strip near the ends, at AB or CD.

The dotted lines are sometimes referred to as *lines of discontinuity*. With the discontinuity lines shown in Fig. 4.18, no two adjacent beam strips will have similar loading, and this implies continuously variable reinforcement patterns.

A more practical choice of discontinuity lines is to step them so as to give finite width beam strips which have similar loading. Yet another alternative available is to suppose the total loading (p) to be split locally into two components, αp and

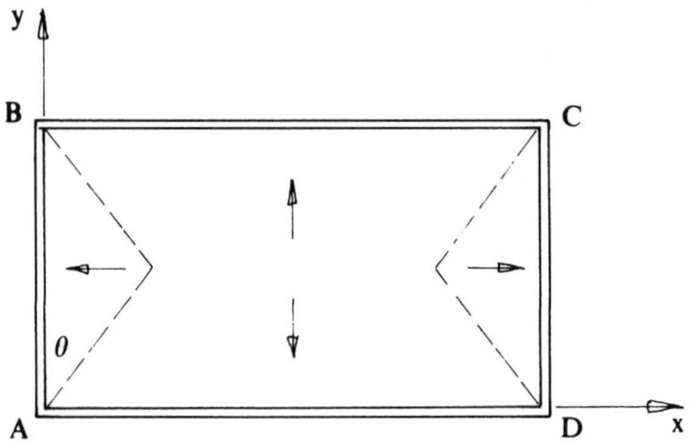

Figure 4.18 Strip theory—spanning proposals.

p(1−α), which are applied to the beam strips in say x and y directions respectively. The simplest such scheme is to take α to be 1 or 0 as is implicit in Fig. 4.18. A more elaborate variant is to think of α as α (x, y) which can be adjusted across the slab for nearness to a boundary, for example.

In actual use, a balance must be struck between simplicity and useful elaboration. Also, for practical purposes there are limits beyond which the strip solution is likely to produce unacceptable cracking in a real construction. For example, in Fig. 4.18 if θ is assumed too small then the y spanning strips will be provided near to the short edges and cracking along these edges will occur in a real construction.

Hence unlike the other methods described, the strip method has an important element of discretion, or choice, associated with it. To the inexperienced this signals caution. To the more experienced this method provides a malleable method which can be made to cope with the unusual situation.

The content of Chapter 5 can be thought of as a systematic development of this theme of strip methods.

III PLATES AND SLABS—THE COMPARISON METHOD AND LOWER BOUNDS

4.11 The Comparison Method–general principles

What has gone before

Non-dimensional groups, or numbers, are useful in plate bending mechanics and have already been introduced earlier in this chapter. This concept is now explored in a more systematic way.

The earlier parts of this chapter deal with plastic bending of plates from the standpoint of considering how the plate *deforms* as it collapses by means of a pattern of yield lines. Deciding on an appropriate *mechanism* of collapse is central to this study. The mechanism chosen must be a possible mechanism. In unfamiliar cases this may not be easy to establish. Any result obtained will be incorrect unless the layout of yield lines chosen does indeed describe a possible manner in which the plate could deform. As a supplement to the checking it could even be useful to make a paper model of the proposed mechanism!

Usually the choice of mechanism is based on experience rather than systematic analysis of the conditions. The calculation then proceeds by computing quantities such as the moment work absorbed in the yield lines and the loss of potential energy of the load. When equated these two quantities ensure equilibrium and lead to an estimate of the collapse load. As has been discussed in Section 4.2, this estimate of the collapse load is an *upper bound* (an over-estimate) of the actual collapse load. Hence it is an *unsafe estimate*—the value calculated is *greater* than the true collapse load. Never-the-less upper bounds are very useful, and are widely discussed in the literature. This is also the most common method used to study plate collapse conditions. The reason for this is that *lower bounds*, safe estimates, or better still, *exact solutions*, are very difficult, if not impossible, to obtain by the commonly available methods.

Recently progress has been made with another, quite different approach to solving plate plastic collapse load problems. This new approach can deal with both upper and lower bounds by a single method and will be called the **Comparison Method**. The idea we exploit is very simple, and very powerful. It will be presented here from a physical rather than a mathematical point of view. References are given in the Bibliography, Chapter 7, where proofs of the statements can be found.

The new concept

The core concept of the method is that we make a *comparison* between the plate being studied and an *equal area circular* plate. All other features are required to be the same: so same edge condition, same thickness, moment of resistance, self-weight etc. For simplicity only *uniformly loaded*, constant thickness, edge-supported, isotropic plates will be considered for the present. These restrictions can be relaxed and later we shall do this. The observation (the comparison) is then that the *equal area circular plate will support less load at collapse than the particular (non circular) plate being studied*. The circular plate is thus the *weakest* shape of plate from amongst all edge supported plates of given area, the other features, self weight etc. being the same for all the plates. This is the core observation on which the Comparison Method depends. Then we make use of the knowledge that the circular plate problems can be solved *exactly* to provide quantitative estimates for the collapse loads of plates of any shape.

In Section 4.4 we studied the collapse features for isotropic circular plates under uniform and other loadings, for both simply supported and clamped edge conditions, and for two yield criteria, the so called Square yield locus and the Tresca locus. For all of these cases we were able to solve the problems *exactly*. It is because exact solutions are so rare that this knowledge of the circular plate exact solutions has provided the motivation for exploitation of these (exact) solutions in the present Comparison Method.

What we found in Section 4.4 was that for the Square yield locus and uniform loading, the collapse pressure, p_c, for a circular, isotropic plate is given by

$$\frac{p_c A}{m_p} = 3k\pi$$

where A=plate area = πR^2 and with
$k = 2$ for simple support and $k = 4$ for a clamped (fixed) edge.

Non-dimensional load—γ—the "weak" bound

Expressed in symbols, and defining the non-dimensional parameter γ, the Comparison statement described above can then be written

$$\gamma \equiv \frac{p_c A}{m_p} > 3k\pi. \tag{4.11.0}$$

Here we are estimating the collapse pressure for the plate shape of our choice, by comparing it with the known exact value for the equal area circular shape.

The inequality (4.11.0) becomes an *equality* if, and only if, the shape of the plate is circular. The plate plan area, A, is seen to be of primary importance and is clearly an easily evaluated quantity. This feature of the prime importance of the plan area, A, is not well recognized in the literature. The non-dimensional quantity, **γ**, is a very convenient parameter to measure the collapse pressure, p_c, of the plate being studied, though as we shall see it is not the only useful parameter, nor always the best choice.

This non-dimensional number γ provides a *lower bound on the collapse load.* This is because the numerical value for γ is always less than the actual collapse value for the particular shape of plate being studied. When we refer to this bound we shall call it the "weak" bound. Often the actual plate collapse mechanism does not spread to the entire area of the plate, and if by some means an estimate of the reduced area of the mechanism is known to us, then this can be used in (4.11.0) to give a better (larger) value for the lower bound. For many practical shapes, such as most rectangular shapes, where the mechanism is known not to extend into the corners then the mechanism will be confined to an area smaller than the gross area of the plate and so an improved bound can usually be found. We shall consider later how this reduced area can be calculated or estimated.

Inequalities such as (4.11.0) are known as *isoperimetric* inequalities, and are quite commonly met with in some parts of mathematics. Later we shall meet other inequalities which look very similar but are not *isoperimetric*, since they do not meet the "if, and only if" condition applying to the circular shape. For our present purposes however this is not vital to their usefulness. For these other inequalities the associated equality condition may apply to other shapes as well as the circular shape! We shall return to this topic later.

Another non-dimensional parameter of at least equal importance alongside $\boldsymbol{\gamma}$ in our present studies is a purely geometrical parameter that relates the plan area (A) of the plate to the length (perimeter) of the boundary (B) enclosing this area. We shall denote this second parameter by the Greek letter $\boldsymbol{\alpha}$. It has been appreciated from ancient times that there is an (isoperimetric) inequality connecting these two quantities, area and perimeter, namely

$$\alpha \equiv \frac{B^2}{A} > 4\pi. \tag{4.11.1}$$

Here again the inequality becomes an equality if, and only if, the shape of the boundary is circular and so is isoperimetric. We say "appreciated" rather than "known" in describing this new parameter because the full content of this inequality and the associated isoperimetric property has only been finally proven in modern times: the truth of the result has been assumed for hundreds of years. In other words (4.11.1) was for long assumed to be true though lacking a proof: it was a *conjectured* result.

Plate area and shape—β—the "strong" bound

These two parameters are very useful, but an even more useful parameter, defined to be β, is the ratio of the previous two: thus

$$\beta \equiv \frac{\gamma}{\alpha}. \tag{4.11.2}$$

We have already two isoperimetric inequalities, (4.11.0), (4.11.1), and it is very unusual for the ratio of two inequalities to itself be an inequality. However in this case this is the situation although the resulting inequality is not of isoperimetric form since there are plate shapes other than circular for which the inequality reduces to an equality.

We now state, without proof, that the parameter β has the property

$$\beta \equiv \frac{\gamma}{\alpha} = \frac{p_c A^2}{m_p B^2} > 0.75k. \qquad (4.11.3)$$

Here the left hand sides have been divided, as have the right-hand sides, and k = 2 or 4, for the simple or fixed boundary, respectively.

The underpinning proof of this statement is discussed in the references noted in Chapter 7, Bibliography and exercises. For the remainder of the present discussion the validity of this result will be assumed. We could refer to this β property as the *strong* bound where the γ property has already been referred to as the weak bound. There is some evidence to suggest that the *Strong* bound may not hold true universally, or for other, more elaborate yield conditions. The *weak* bound does still apply for these other yield conditions such as the Mises condition. If some of the bound properties do prove to be violated for these other yield regimes, the relations such as (4.11.3) are likely still to provide upper bound collapse load information, much as yield line theory does. The big difference between the present Comparison Method and alternatives such as yield line methods is the dominant role geometry plays with the Comparison Method. Geometry has a role in yield line but it is no more dominant than the role of equilibrium. Equilibrium is considered in the Comparison Method but it is explicit only in the mechanics of the circular plate as the core comparison.

At this stage we should consider more carefully the precise meanings attaching to A and B. We have described A as the "plan area" of the plate. Recall that we are investigating the collapse conditions for plates and as the plate collapses a mechanism of collapse will form. Then the precise meaning of A is that it is the *area which spans (covers) the mechanism of collapse* for the plate being studied. In the case of the circular plate the mechanism is known to extend to the outer supported boundary and hence A is the total area of the plate. For other shapes, especially the more commonly met plate shapes such as rectangles and shapes with corners, the mechanism may not extend out to the entire edge of the plate, particularly in the corners for example. Then A is the (lesser) area of that part of the plate that takes part in the collapse. In the corners of say a rectangular plate, the converging edges provide enhanced support and hence close to the corner the plate is stronger than the surrounding parts of the plate. The mechanism then you might say, is repelled and forms in the plate away from the corner. These sorts of mechanism have already been encountered in Section 4.7 for example. It is always the case that B is the boundary length, the perimeter, of A. For most shapes of plate at collapse both A and B are unknown and their values will be part of the solution, since it is not known at the outset how far the collapse mechanism extends.

An example

In Section 4.7 the yield line method was applied to the simply supported rectangular shape, with side lengths L and l, uniformly loaded, isotropic plate. Let the plate bending strength/unit length be denoted by m_p, and the collapse pressure by p_c. It is p_c that we are trying to calculate: m_p is part of the given data. Then (4.11.3) can be applied. First we must decide what is known, or can be assumed, about A and B. Here we shall make the same assumptions, which are also the simplest, as are made in Section 4.7—namely that $A = Ll$ and $B = 2(L + l)$. With these values it is being assumed that the mechanism of collapse extends over the entire plate area and to the entire plate boundary. As will be seen later, this is a simplifying assumption and leads to inevitable inaccuracy in the calculated value for p.

Using these values in (4.11.3) and rearranging we obtain

$$\frac{p_c Ll}{m_p} > \frac{6(L+l)^2}{(Ll)} \tag{4.11.4}$$

As the ratio L/l increases from $L/l = 1$, so the (rectangular) shape elongates. When we compare values obtained from (4.11.4) with those from Table 4.0 we see that the approximate values for p_c obtained by the present (Comparison) method are consistently *less* than those obtained by the yield line approach. In the case $L/l = 1$ the two methods give the same value of 24 for the non-dimensional ratio, pl^2/m_p. But the present values, though less than, or equal to, the yield line values, are not lower bounds on the collapse load, p_c. They are just better upper bounds than those obtained earlier by the yield line method. Even better upper bounds can be obtained by the Comparison Method if more realistic choices are made for A and B, just as has been the experience with the yield line approach.

The relationship (4.11.3) is a very simple and powerful expression for obtaining estimates of the collapse pressure for *any shape* of isotropic, edge-supported, uniformly loaded, ductile plate. A restriction is that there must be support at the edge, and it must be of the same type at all points around the boundary. So it must be either a simple support or a fixed (clamped) edge support. Free edges are excluded for now. We shall consider examples of other shapes and edge conditions later: see Section 4.15.

4.12 The Comparison Method—lower bounds on the collapse load

Upper bound expressions of the type (4.11.4) are very useful, but our aim is to give more precise meaning to the value of the collapse loading intensity, p_c.

We would prefer to have a *lower bound* for p_c, and *equality* rather than an inequality as the working relationship.

As has been remarked on already (Section 4.11), if we stay with the bound theorems as described in Section 4.2, then, in principle, lower bounds could be found if an approximate solution is sought in which *equilibrium and yield* are both satisfied *at all points of the plate*. These requirements are easily stated but very difficult to satisfy and at the same time produce useful lower bounds. Even in those cases where lower bounds have been found by this approach, the results are often not useful because the bound is too far below the exact value. Thus the situation is that although the requirements for finding lower bounds on p_c have been known for at least fifty years, very few "accurate", useable lower bounds have been found by limit analysis based procedures. Because of the practical inability of limit analysis methods to provide useful lower bound values, so the Comparison Method has been developed as an alternative approach which can often provide very accurate lower bound solutions for a wide range of plate collapse situations.

We now describe the means by which the Comparison Method is able to produce lower bounds on p_c. These are comparatively easily computed, and are often surprisingly accurate in the sense that these lower bound values are very close to the exact values where these are known.

The expression (4.11.3) will first be rearranged to the form

$$\frac{p_c}{m_p} > 0.75k\left(\frac{B}{A}\right)^2 \qquad (4.12.0)$$

Next the search will be made for the *shape* of the deforming zone in the collapsing plate which *minimizes* the ratio B/A! If we can find this $(B/A)_{min}$, and if an equality replaces the inequality in (4.12.0), then the estimate of value for p_c obtained from the present calculation is a *lower bound* on p_c, and will be denoted by p_L. Hence we have

$$\frac{p_L}{M_p} = 0.75\left(\frac{B}{A}\right)^2_{min}. \qquad (4.12.1)$$

This expression is the key relationship of the present lower bound studies. Here, as earlier, k=2 for simple support and k=4 for a fixed edge.

Investigation of the B and A for the given case to give a minimum for the ratio B/A is a purely geometrical problem, and as such is in principle much easier to deal with than many of the other tasks encountered in the analysis of plates. Even so it turns out that this is not a trivial problem. Fortunately this question has been studied by mathematicians over a lengthy period. The stage that they have reached does not exhaust the possibilities, but their interim conclusions are useful in this study.

We shall not give a proof of the result, but the search for the B and A to give a minimum B/A comes down to a quite simple search. If the plate shape is smooth, such as the elliptical or circular shapes, then the mechanism of collapse generally extends to the whole area of the plate and the relevant B/A is then obtained by using the whole area, **A**, and perimeter, **B**, in the expression (4.12.1). Most practical plate shapes have corners and straight edges: typically square and rectangular shapes. Then what we have to do is draw inside the actual plate boundary a slightly *smaller* shape, indeed a shape with rounded corners inside which the mechanism of collapse forms. This shape is inscribed in the given shape to lie along the actual plate boundary except near the corners. In the corners a circular arc, of radius r say, where r is an unknown, must be drawn to be tangential to the two edges bounding the corner, and joining therefore smoothly with the rest of the boundary which lies along the plate edges. See Figure 4.19. There is no information about the details of the mechanism which forms inside this inner boundary. But this is not really a disadvantage. Even if we had the detailed information, it is not required for use in any part of our lower bound solution. Nor is there any need to have the information for later design purposes, since the plate is isotropic and hence of equal strength in all directions and at all points across the plate.

What we are left with finding is the value of this all important corner radius, r. Once we know r then we can complete the drawing of the whole of the boundary, B, of the deforming zone. Knowing B we can compute the area, A, which covers the collapsing part of the plate. Finally we have the quantity we are seeking, namely the *lower bound estimate for the collapse load,* p_L from (4.12.1). What we shall find is that there is a stunningly simple relationship between the lower bound collapse pressure and this corner radius!

It is relevant to point out that we cannot give a fully comprehensive proof of this result for finding the minimum value of B/A and for a general shape of boundary. Strictly, only shapes for which an in-circle can be drawn are proven to lead to the minimum of B/A by the methods we are using. Even so there are strong physical reasons to suppose that this procedure does lead to the minimum of B/A for shapes with no in-circle. In practical cases most shapes will fall into this latter category. We shall *conjecture* the result to be true for any cases where the mathematical proof is lacking. Indeed this is another reason for discussing the subject from this viewpoint. There is scope for young workers and researchers to progress the subject by studying these sorts of questions and aiming to construct proofs where we are still relying on conjectures to provide working solutions to such topics as lower bounds on the collapse loads.

4.13 Finding the r_{min}—a geometrical problem

Suppose we have a plate of total area **A** and boundary length **B**. The shape we shall suppose to be some general polygon. What we are seeking to find is a *smaller* inscribed area, A with boundary length denoted by B, where the area A extends to almost all of the original plate area and encloses the whole of the plastically deforming zone when the plate is loaded to collapse. We expect this deforming zone to cover most of the plate, except in the convex corners of the plate where the strength deriving from the edges prevents the mechanism from extending into the corner. A convex corner encloses an angle <180 degrees when viewed from plate interior. If the corner encloses an angle of more than 180 degrees we shall refer to such a corner as a re-entrant corner. Then the boundary B (of A) will lie along the original boundary everywhere except close to these convex corners. At a re-entrant corner the deforming zone is likely to extend right up to the edges and there will be no "dead zone" of rigid material in the corner. Instead of extending fully into the convex corner the B-boundary follows a circular arc of radius r, and this happens at all convex corners. It is the value of this common corner radius, as yet unknown, that we are seeking to establish. These corner circular arcs and the plate edges must meet smoothly and hence the boundary B will be smooth everywhere except at any re-entrant (concave) corners in the plate. There the B-boundary will hug the original plate boundary. In addition to all of the above, we shall seek to find the A and B shapes such that their ratio shall be a *local minimum*. It is this condition which produces the value for the radius r, which we shall refer to as r_{min}.

Mathematically speaking the above procedure is known to produce the $(B/A)_{min}$, but only for a restricted range of (plate) shapes. These are the shapes that possess an *in-circle*. Such shapes represent only a tiny number of the plate shapes we wish to study. For example, all rectangular shapes, other than the square, do not possess in-circles, and we certainly wish to study these shapes. In what follows we shall *assume* that the above (mathematical) procedure, applicable to shapes which have an in-circle, will continue to apply for other more general shapes as well, though there may be pathological cases in which this is not true. Then in order to find $(B/A)_{min}$ we will expect the corner radius, r, to be present in all convex corners, and with the same radius for all corners in the particular plate. In addition we shall continue to apply the same criteria to any shape, with or without an in-circle, and whether or not the shape is convex or concave. As we shall find later, even when pathological cases arise, the information which expressions like (4.12.1) provide is valuable and generally quite usable.

Hence there may be some situations in which the methods break down, but usually this will become evident from the value of r we compute, and from other evidence. The shape shown in Fig. 4.19 is an example of a shape where there is one re-entrant corner, which means that the shape is non-convex. We shall return to this shape later.

But to return to our general polygonal shape of plate! Such a shape is formed from a series of straight sides which meet in pairs to form the polygonal shape. The shape can be described by the number of exterior (convex) corners, denoted by e_α, and the number of internal (concave) corners denoted by i_α. For the shape shown in Fig. 4.19, $e_\alpha = 5$ and $i_\alpha = 1$.

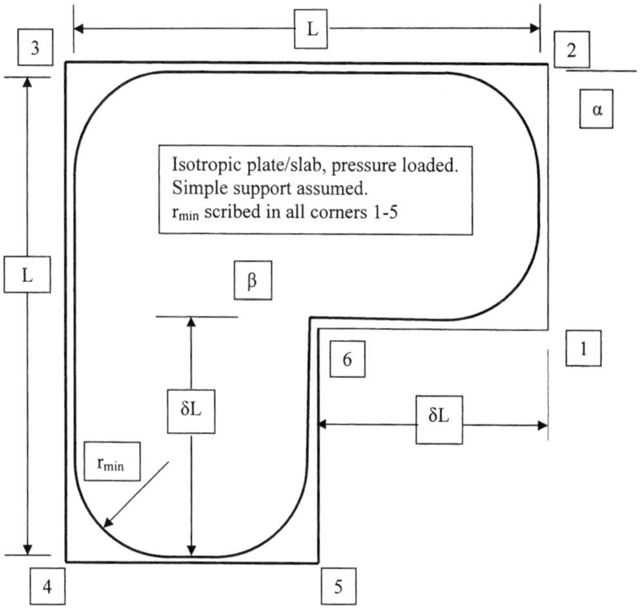

Figure 4.19 A typical non-convex shape of plate (Figure drawn for $\delta = 0.5$, then $r_{min} = 0.198.L$ and $p_L.L^2/m_p = 1.5.L^2/r^2_{min} = 38.25$. If $\delta = 0.75$, then $r_{min} = 0.113.L$ and $p_L.L^2/m_p = 117.92$).

Exterior (convex) corners, such as corners 1–5 in Fig. 4.19, will be described by the *external* angle, a_α—see corner 2 in Figure 4.19. At an interior (concave) corner, such as 6 in Figure 4.19, the *internal* angle ϕ_β will be used to describe the shape of the corner.

Then it is straightforward to show that for a general polygonal shape with α convex and β concave corners

$$\Sigma\, a_\alpha = 2\pi + \Sigma\, (\phi_\beta - \pi), \quad \alpha + \beta = \text{number of vertices} \qquad (4.13.0)$$

where each vertex contributes one term to the equation.
Exercise: In Fig. 4.19 $e_\alpha = 5$, $i_\alpha = 1$, all five $a = \pi/2$ and $\phi_\beta = 1.5\pi$. Using these values (4.13.0) is seen to be satisfied.

As noted earlier, the total area of the plate will be denoted by **A** and the boundary length of this area by **B**. With A denoting the area of the deforming part, as earlier, and B the boundary length of A, then

$$A = \mathbf{A} - r^2.\omega \qquad (4.13.1)$$

$$B = \mathbf{B} - 2r\omega \qquad (4.13.2)$$

where

$$\omega \equiv \Sigma \tan(a_\alpha/2) - \Sigma\, (a_\alpha/2) \qquad (4.13.3)$$

and the α's are summed over the *exterior* (convex) corners only.

Now (B/A) is a function of r, and hence for the ratio to be a minimum the derivative with respect to r must be zero. Hence we require that

$$A \, . \, \frac{dB}{dr} = B \, . \, \frac{dA}{dr} \, . \qquad (4.13.4)$$

When we substitute the values for A and B from (4.13.1) and (4.13.2) into (4.13.4) the resulting expression is quadratic in r_{min} and is the important relation from which r_{min} is obtained.
Thus

$$\omega r_{min}{}^2 - \mathbf{B} r_{min} + \mathbf{A} = 0. \qquad (4.13.5)$$

The useful solution is

$$r_{min} = \mathbf{B} - \sqrt{(\mathbf{B}^2 - 4\omega\mathbf{A})}/2\omega. \qquad (4.13.6)$$

But we note from (4.13.1) and (4.13.2) that

$$r_{min}(\mathbf{B} - 2r_{min}\omega) = \mathbf{A} + \omega r_{min}{}^2 - 2\omega r_{min}{}^2 = \mathbf{A} - \omega r_{min}{}^2.$$

Hence we see that

$$\left(\frac{B}{A}\right)_{min} = \left(\frac{1}{r}\right)_{min} \qquad (4.13.7)$$

This remarkably simple relation, when used in (4.12.1), then gives us the most significant and useful relationship of the present study, namely

$$\frac{p_L}{m_p} = 0.75 \ k/r_{min}^2 \, . \qquad (4.13.8)$$

Remembering that the constant k takes the value 2 for a simply supported edge and 4 for the clamped edge, we can now, by use of (4.13.8), find lower bound values for the collapse load, p_c, for plates of arbitrary shape.
Exercise: The most significant shape of plate to consider at this stage is the square of side length L under pressure loading.

Here all the $a_\alpha = \pi/2$, for all four external convex corners. There are no re-entrant corners and so $\phi_\beta = 0$.

Therefore

$$\omega = 4\tan(\pi/4) - 4\pi/4 = 4 - \pi,$$

also $$\mathbf{A} = L^2 \quad \text{and} \quad \mathbf{B} = 4L.$$

Substituting into (4.13.5), the quadratic for r_{min} is given by

$$(4 - \pi)\, r_{min}^{\,2} - 4L\, r_{min} + L^2 = 0$$

from which we find

$$\frac{r_{min}}{L} = \frac{(2 - \sqrt{\pi})}{4 - \pi} = 0.2651.$$

Finally,

$$\frac{p_L}{m_p} = 0.75\ k/r_{min}^2. \qquad (4.13.8)$$

with k = 4 for the clamped edge plate. Then the collapse pressure is given by

$$\frac{p_L L^2}{m_p} = 42.694. \qquad (4.13.9)$$

This value compares with the known exact solution of 42.851: see Section 7.2 Exercises (3).

If instead of the square a general rectangular shape is considered, the quantity omega, ω, remains unchanged at $4 - \pi$ but the value of r_{min} changes because the side ratio is now not unity. If the sides of the rectangle are L and l, then $\mathbf{A} = Ll$ and $\mathbf{B} = 2(L + l)$. Using these values in (4.13.5) for the minimum corner radius, r_{min}, the lower bound values for the rectangular shape are obtained.

Exercise (1) Generalize the result for the square to any regular n-sided polygonal shape. If the side length is L, show that $\omega = n\tan(\alpha/2) - \pi$, $\mathbf{A} = (nL^2 \cot(\pi/n))/4$, $\mathbf{B} = nL$, where $\alpha = 2\pi/n$, and hence r_{min} follows from (4.13.5). As n increases so the circular shape is approached, with $pA/m = 3k\pi$ in the limit, k = 2 for simple support, k = 4 for fixed edge.

Exercise (2) Consider a clamped rhombic plate. This is the four-sided shape of four equal length sides of a square shape which has been deformed by lengthening one diagonal and hence shortening the other. If we define the acute angle in a vertex to be β, then if the side length is L and the radius of the in-circle is a, we have $a = L\sin\beta/2$, $\mathbf{A} = L^2\sin\beta$, $\mathbf{B} = 4L$, $A = \mathbf{A} - 2r^2(T + 1/T) + \pi r^2$ and $B = \mathbf{B} - 4r(T + 1/T) + 2\pi r$. Here we have defined $T = \tan(\beta/2)$. As earlier **A**, **B** are the gross area and its boundary length; A, B are the smaller area and its boundary length, inscribed in the gross area and inside which the collapse mechanism forms. For the case $\beta = \pi/2$, the square, find the value of r to minimize the ratio B/A. Show this to be given in general by $r/a = 2(2 - \sqrt{(\pi - \sin\beta)})/(4 - \pi\sin\beta) = 0.5302$ for $\beta = \pi/2$ and thence the lower bound collapse pressure to be $p_L a^2/m_p = 10.67$. (The exact solution for this shape, the square, is known to be 10.71. See the Exercises in chapter 7.) We commented earlier that the Comparison Method spans both upper and lower bounds in a continuous gradation. This can be illustrated in this example by giving the corner radius r any reasonable value and then the collapse pressure obtained will now be greater than the lower bound found by the above methods: it could still be a lower bound but will more likely be an upper bound. When we draw a graph for values $0 < r/a < 1$ this is a gently curving plot with a minimum at $r/a = 0.5302$ and giving a lower bound of 10.67. The axes are pa^2/m_p and r/a and the coordinates of the end points are 12, 0 and 12, 1. These same end point values are produced by the yield line upper bound approach using a simple pyramid and a conical yield mechanism, respectively. The best yield line can do is a least upper bound.

Exercise (3) Refer again to Fig. 4.19. If the shape is square, side length L and with a smaller square shape cut from it of side length δL, $0 < \delta < 1$, consider the edges to be supported and either fixed or simply supported. The loading will be taken to be a uniform pressure of p/unit area. Now when we seek lower bounds on the collapse load, we need to find the radius, r, of the corner portions of the deforming zone as earlier. For small values of δ some of these corner zones may not subtend angles of $\pi/2$ at the circle centre despite the corner being itself right-angled. This is what you might call a pathological case. We can still solve for r_{min} but it is a little more difficult to write down the expressions for A and B. As an exercise let us find the largest value of δ such that all the corner radii do subtend angles of $\pi/2$ at the respective circle centres. Show that the quadratic for r is $(4 - 1.25\pi)\,r^2 - 4Lr + L^2 = 0$ and this solves to give $r/L = \delta/L = 0.2512$. This is the smallest value of δ to satisfy the conditions. Then for the fixed edge, the lower bound non-dimensional parameter $pL^2/m = 47.56$. An upper bound for the same shape of boundary (namely $\delta = 0.2512L$) is based on the simplest choice of mechanism that extends into all the corners with $r = 0$. Then $A = \mathbf{A} = (1 - \delta^2).L^2$ and $B = \mathbf{B} = 4L$ and the non-dimensional parameter $p_uL^2/m = 3L^2(B/A)^2 = 54.68$.

4.14 Affinity Theorem—orthotropic plates—associated isotropic equivalents

In practice reinforced concrete plates (slabs) are frequently orthotropic—meaning that the amounts of reinforcement are not equal in the two primary directions. Usually these two primary directions are parallel to the supported edges, for a rectangular shape of plate for example. There may also be both top and bottom steel, and these amounts may be different. See Chapter 6 on construction and design for further explanation of these features.

The Comparison Method arises naturally in the context of isotropy where there are no preferred directions in the plate. The Affinity Theorem is a means of relating an orthotropic plate to an associated isotropic equivalent. Once the isotropic equivalent is known then the Comparison Method, and other methods, can be applied to find upper and lower bounds for the collapse conditions.

Consider again the rectangular plate. We shall think of it as a reinforced concrete construction, with unequal amounts of reinforcement in the two directions parallel to the edges. Hence it will be orthotropic. We have previously considered such an example, in Section 4.9. Suppose that the area of reinforcement running in one direction is μ times that running in the other direction. This is what we have already termed the orthotropic ratio. Call these two directions x_1 and x_2. Let the length of the plate in these two directions be L_1 and L_2 with the 1 direction along the shorter side.

The Affinity Theorem we wish to use states that for a plate with orthotropic ratio μ, such that the plate strength $M_2 = \mu M_1$, we can associate an isotropic plate with strength M_1 and a shape derived from the orthotropic plate shape such that lengths measured in the x_2 direction are scaled by the factor $1/\sqrt{\mu}$. The length L_1 is unchanged. These two plates the theorem states have the same collapse (pressure) load, p. There are other aspects to the theorem which the reader can find in the books by Johansen, Nielsen and Wood, listed in the Bibliography.

This theorem, used in conjunction with the Comparison Method, is then able to provide both upper and lower values for the collapse load for edge supported, orthotropic plates of arbitrary shape. Let us return to the rectangular orthotropic plate with ratio μ. For the shape with long side twice the short side length, l, and the short side being direction 1, then the long side is direction 2 and the length is scaled to be $2l/\sqrt{\mu}$. The scaled shape is now of an isotropic slab of uniform strength M_1, boundary length $\mathbf{B} = 2l(1 + 2/\sqrt{\mu})$ and $\mathbf{A} = 2l^2/\sqrt{\mu}$. These values used in (4.13.5) then give the value of r_{min}. The collapse load follows from (4.13.8). If $\mu=1$, the isotropic case, then pA/m = 24.36 where m is the isotropic strength for a simply supported plate. If $\mu = 0.2$, in which case most the reinforcement is spanning the short dimension, l, then pA/m = 28.00, with $M_1 = 2m/(1 + \mu)$ and $M_2 = \mu M_1$. Changing to an orthotropic arrangement gives a strength increase for the same total amount of reinforcement. There is also a value of μ that gives a maximum strength. For the 2 by 1 rectangle this value is about 0.18: the benefit is a small additional increase in collapse load.

Had a μ value greater than one been chosen, this would mean more steel in the longer dimension of the plate than across the short dimension. The strength would be less for the same amount of steel, because steel in this longer direction is less effective. We have already observed this in Section 4.9. It is commonly held that the μ value should not be too extreme, meaning too far away from the isotropic, $\mu = 1$ value. This view is largely based on accumulated experience with actual slabs over the years. Directions in thin concrete members such as slabs which are very lightly reinforced do tend to develop cracks across these directions. Much of this cracking is the result of shrinkage of the concrete, which does occur as a natural process, though it can be minimized by careful curing of the concrete. Any steel present which crosses the crack direction will help reduce such cracking. Such steel is termed "distribution steel" if it is not placed there primarily for strength requirements.

One of the common features of the solutions to the optimal slabs examples considered in Chapter 5 is that over much of the area of the slab only one direction, that of the constant principal curvature, requires reinforcement for strength requirements. All these layouts thus come into the category of advocating an extreme value of μ namely $\mu = 0$. This is probably one reason why these layouts, though known about for decades, have attracted very little interest from the profession or industry. Another is that the reinforcement requirements vary from point to point across the slab. Such variation cannot be achieved in practical construction. The optimal studies do have merit however. They are one of the few models of structural behaviour that produce a "design" in the sense of specific detail about the strength (hence actual reinforcement amount) needed throughout the slab. If we do not like what the theory/design tells us because it is too extreme or too difficult to build or for other reasons, so be it. The real benefits from considering the slab structures from this viewpoint are the *trends* that the solution points to. From a simple summing of some numbers we are told how much steel of the total is needed in a particular zone of the slab. We need not use the reduced amounts that the optimal theory says are the optimum, but we will probably not improve on the optimal theory data for how to distribute what reinforcement is finally employed.

Typically the optimal theory consistently points to the corner zones in edge supported slabs as requiring only small amounts of reinforcement. This is counter to much established practice. Elastic theory of plates often indicates singularities in corners. This in turn tends to focus attention on the corners. But in reality corners of edge supported plates are close to supports and should not attract reinforcement. The situation will change if the edge support is reduced or removed. Then corners may attract major moment concentrations and hence reinforcement requirements.

4.15 Other edge conditions

All the plates we have considered thus far have been fully edge supported. For the Comparison Method to function most efficiently the edge must be fully edge supported and with the same edge condition around the entire boundary. Other edge conditions can be dealt with but in an approximate manner.

An unsupported or free edge can be accommodated approximately by the Comparison Method and will lead to an upper bound collapse load estimate. Our approach is to think of the actual plate with a free edge and then reflect the plate in the free edge to give a plate of twice the size and the free edge as a line of symmetry in the centre of the modified plate. Then the reasoning will be to analyse this enlarged plate, with the free edge a line of symmetry. This modified plate is of course now supported around the entire boundary though our real interest is only on half of it and it is intended to be free on the remaining side that is the line of symmetry. The values found will be upper, unsafe bounds on the collapse load since the actual free edge as treated here will in general have restraining edge moments but there will be no shears because of the symmetry. Hence one of the two free edge conditions is exactly satisfied, but the moment condition in general is not. A similar situation arises in yield line analysis where the free edge case has the same collapse load as the doubled-in-size plate. These are all situations in which the transverse slope at the edge is zero.

Consider the rectangle of side lengths L and l with $L > l$ as considered earlier, but now free on one of the shorter l-length edges. The remaining edges may be either simply supported or clamped. Analyse for a uniform pressure loading, p/unit area. Using the simplest mechanism of a single yield line into each supported corner, and equally inclined to the edges, then the yield lines joining with a yield lines out to the centre of the free edge, the estimate becomes

$$\frac{pl^2}{m} = \frac{24(2L + l)}{(6L - l)} \tag{4.15.0}$$

for simple support.

The crudest approximation by the Comparison Method is obtained with $\mathbf{B} = 2(2L + l)$ and $\mathbf{A} = 2Ll$ and we obtain

$$\frac{pl^2}{m} > 0.75k(B/A)^2 = 1.5(2 + l/L)^2 \tag{4.15.1}$$

for simple support, $k = 2$.

The comparison method result (4.15.1) gives a lesser value and hence a better upper bound for the collapse load than this yield line estimate, for all $L/l > 0.5$. A different yield line mechanism is needed for $L/l < 0.5$. The results are likely to be poor for such side ratios. These upper bound values can be improved by using a more comprehensive mechanism for the yield line and by seeking a $(B/A)_{min}$ value for the comparison estimate. Now the comparison solution will not be a lower bound, merely a better upper bound.

The treatment of mixed edge conditions and various other as yet unexplored features of the Comparison Method could provide worthwhile topics for further investigation. Readers might take up some of these challenges.

4.16 Conclusions

In a number of respects, the contents of this chapter, more than any other, are central to current plate problems. Certainly, more progress can be made with many of the problems posed here than is the case with the majority of elastic problems. Some of the more general statements concerned with the non-dimensional ratios offer worthwhile distillations and begin to provide a useful categorization of solutions.

In the next chapter some more modern and radical theory is outlined. This theory may be a growing point in the subject in future, though only time will tell.

5 Optimal plates

5.0 Introduction

The emphasis in the previous chapters has been on the study of isotropic plates. As very many practical plates are isotropic, the emphasis on isotropy is reasonable, but there are also valid reasons for making a separate study of anisotropy, which is the central theme of this chapter.

From the practical point of view, all the plates studied in this chapter should be thought of as reinforced concrete plates (slabs), since the anisotropy is physically most easily introduced by varying the amount of reinforcement in chosen directions in the plate.

An independent reason why attention should be paid to the methods discussed in this chapter is that they can be used to solve, in closed form, a technically interesting range of anisotropic plate problems. It will already be evident that many interesting and apparently simple isotropic plate problems cannot be comprehensively solved by the classical methods of Chapters 3 and 4. Hence any (non-classical) method of formulation and solution which allows many closed form solutions to be obtained merits some study. This is the situation with *optimum plate theory* studied in this chapter.

5.1 Problem formulation

The basic aim is to investigate layouts of reinforcement for reinforced concrete, or other fibre-reinforced composites, such that the amount of reinforcement needed to support the load is a *minimum*, or near minimum. The plates will be assumed to be of constant thickness and loaded with static loads. The bulk of the plate will be composed of the matrix (concrete or other filler), reinforced in the *tension* face of the plate with steel reinforcement (or other fibre such as glass or carbon).

The purpose of the study is to suggest amounts of reinforcement, and where and in what directions these should be placed, to achieve minimum fibre consumption for given load. Alternatively, the theory and methods of solution can be seen to produce maximum load capacity for given fibre content when this is arranged optimally.

Consider first a uniform depth and width concrete beam carrying a uniformly distributed load (p) and simply supported at the ends. The bending moment distribution in this beam is determinate—in this case it is parabolic variation from zero at the ends to a maximum central (sagging) value of $pL^2/8$, where L is the span. Suppose now that only *just* sufficient (tensile) reinforcement is provided in the concrete to support this bending moment. Then with the assumption that the depth of the concrete stress block is *small* compared with the beam depth, the conclusion to be drawn is that the steel force (= stress (σ) × sectional area (A)) must be *proportional* to the bending moment (M). Indeed

$$M = \sigma A d$$

where d is the effective beam lever arm, *assumed constant*. Suppose further that the steel working stress can be held constant at some suitable fraction of the yield stress, then it is concluded that

$$A \propto M.$$

In other words, the cross-sectional area of the (steel) tension reinforcement (A) required is proportional to the local bending moment (M).

Hence the total volume of (steel) reinforcement required to support this bending moment distribution is given by

$$V_R \equiv \int_0^L A \, dx,$$

and

$$V_M \equiv \int_0^L |M| \, dx,$$

where V_R is the volume of reinforcement, and V_M is defined to be the *moment volume* associated. Minimizing the moment volume achieves the same result as minimizing the volume of reinforcement, V_R. In what follows, most attention will be focused on V_M.

In the present example, with the x coordinate measured from one end of the beam,

$$M = \int_0^L |M(x)|\,dx = \int_0^L \left| -\frac{px(L-x)}{2} \right| dx$$
$$= \frac{p}{2}\left[\frac{L^3}{2} - \frac{L^3}{3}\right] = \frac{pL^3}{12}.$$

In order to achieve the condition that σ, the fibre stress, be a constant, it will be evident that the beam *curvature* must be constant, since σ constant implies a constant *strain* in the reinforcement and this can only bc achieved from a constant curvature.

This observation about the *constancy of the beam curvature* is the all-pervading flavour of the treatment of optimum plates which follows in this chapter.

Though full proofs of the necessary and sufficient conditions for achieving a minimum of V_M, and hence V_R, will not be developed, it can be appreciated after a few examples that optimum (or minimum) configuration of curvatures and hence deflected shape, and associated bending moments conform with the following requirements.

First, curvatures and bending moments must *correspond*. By this is meant that sagging curvature must be associated with sagging bending moment. Secondly, the deflected shape must be *smooth*. This means that adjacent constant curvature segments must meet along a common curve and the slope *transverse* to this common curve must be common to the two segments.

Those directions which are reinforced must be directions along which the curvature is *constant*. For if this was not the case, then the reinforcement would not be strained to the constant value desired.

Again the stressed reinforcement determines the internal bending moments and these bending moments must be in equilibrium with the loading. So equilibrium must be achieved, with constant curvatures along the reinforcement directions, and the deflected surface segments must fit together to produce a *smooth* whole. These appear to be very difficult conditions to satisfy simultaneously, with little constructive guidance to assist. But there is considerable scope for intuition to play a part here, as will be demonstrated shortly.

A key element in the search is to *expect* that, very often, the appropriate bending moment system is likely to be *one-dimensional*. The examples will illustrate this observation.

5.2 Constant curvature surfaces and principal directions

All the constant curvature surfaces considered here are characterized by constant curvature trajectories which are *straight* when seen in plan. Recall that the deflected surface of a deformed plate is a shallow shape. This implies that the vertical section taken along a *constant curvature direction* will reveal a *shallow circular arc* of section. When described by a local cartesian coordinate system, such a piece of arc will be parabolic since slopes are small and squares are to be neglected compared to unity.

If the constant curvature has the value k, then point A, a distance d from a zero slope point (B) of a constant curvature arc, will have a transverse displacement

$$w_A = w_B \pm \frac{kd^2}{2}$$

the sign depending upon the sign convention adopted for curvatures and displacements.

Likewise the slopes at A and B will differ by an amount $|kd|$. Both these results are very useful in exploring the constant curvature shapes which will shortly be considered, and are based upon the identity of a *shallow* circular arc with a parabolic curve.

In the Appendix a good deal of attention is paid to the analysis of surface geometry to allow such features as principal directions on surfaces to be analysed. Although principal directions play a vital role in optimum plate studies, the analysis needed to facilitate this study is quite minimal. The essential feature of a principal direction (refer to Section A.6, p. 220) is the property that if two adjacent points on the surface are considered, and the normals to the surface are drawn at these two points, then the *direction* defined by the *join* of these points is a *principal direction* if and only if these two adjacent normals are *coplanar*. Here, consideration is being restricted to *straight* constant curvature directions, and it is these directions which are intended to be principal directions.

A very important theorem will now be stated which identifies when these straight, constant curvature directions will be principal directions on the shallow deflected slab surface. Consider two points (1, 2), not necessarily adjacent, on a straight constant curvature direction. Generally at each of these points the surface in the transverse direction will have some slope, which will be denoted by S_1, S_2 at the two points 1, 2 respectively. The theorem states that if a given straight trajectory on a shallow surface is a *constant curvature principal* direction, then the slope transverse to this direction is the *same* at *all* points. This result follows from the requirement that adjacent normals be coplanar. Consider the use of the theorem in the quest for *principal constant curvature* directions. To be principal, adjacent normals must be coplanar. This requires that the S values at the adjacent points must be the *same*, since otherwise, viewed with the joining piece of arc end-on, the normals would be skew. Now apply the theorem and note that if the S value is the same at *two adjacent* points it is the same at *all* points. Hence a necessary and sufficient condition for a straight shallow constant curvature arc to be a *principal* direction on the surface on which it is scribed is that the value of the transverse slope S at all points be the same. This result is of fundamental importance in determining whether surfaces and directions are suitable for satisfying the optimal requirements.

To summarize then, the constant curvature surfaces of interest in the theory of optimal plates are shallow surfaces possessing straight constant curvature directions which are principal directions on the deflected slab surface. Such directions are characterized by the property that along any particular trajectory the surface slope *transverse* (that is at right angles) to the constant curvature direction is *constant*.

5.3 Basic results—corners

This introduction to optimal plate theory will be presented from an intuitive viewpoint to begin with; then some of the supporting theory will be discussed. Here it is proposed to examine the layout of optimal, constant curvature directions in the neighbourhood of intersecting edges. From these beginnings, whole problems can later be solved.

The guiding consideration in the intuitive, physically motivated study of these problems is that the optimal spanning layout is likely to be *one-dimensional*. In other words, locally, any optimal zone will resemble a series of constant curvature beam elements. This tendency for such layouts to be one-dimensional has been suggested by theory and will be discussed later in a more formal manner.

Once this tendency has been identified, several solutions for corner zones quickly follow. Consider first a piece of slab between two intersecting simply supported edges.

(a) The simply supported corner

The aim is to describe both a possible spanning arrangement and a deflected shape for the slab in constant curvature terms, in the neighbourhood of the simply supported boundaries.

Consider the strip AB drawn normal to the angle bisector. This strip can support any distribution of superimposed load and self weight, by suitable simple beam reinforcement. Also if this beam is bent to a constant sagging curvature of $-k$, then all this reinforcement will be strained to a constant amount and hence to a constant stress level. These features are the ingredients of an optimal system.

While it has not yet been proved it will be evident from considering alternative arrangements that the series of parallel spans, of which AB is a typical one, provide the optimal layout. Taken together the series of parallel spans generate a rather simple constant curvature surface. From symmetry it is true that AB is a principal direction on this surface. Also the second principal curvature, namely

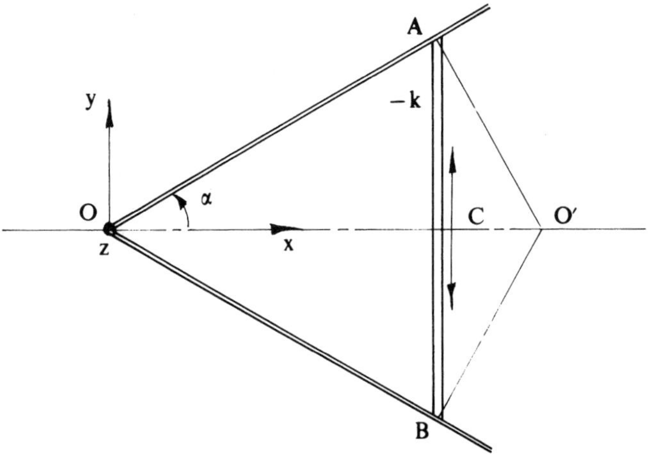

Figure 5.0 Simply supported corner arrangement.

that curvature which is normal to the constant curvature along AB can be evaluated as follows.

Set up a cartesian system of coordinates at the corner (Fig. 5.0). Then with an included angle of 2α,

$$AC = x \tan \alpha.$$

Hence, using the expression from 5.2 for a constant curvature arc, since C is a zero slope point, the difference in transverse displacement between A and C is given by

$$w_c = \frac{k}{2}(AC)^2,$$

since $w_A = 0$ from being on the simply supported edge.

Thus

$$w_c = \frac{k}{2} x^2 \tan^2 \alpha.$$

The second principal curvature, κ_2, is then given by

$$\kappa_2 = \frac{d^2 w}{dx^2} = k \tan^2 \alpha. \tag{5.3.0}$$

From this expression it is seen that $|\kappa_2| \leqslant k$ so long as $\alpha \leqslant \pi/4$. When $\alpha = \pi/4$ then $|\kappa_2|$ is equal to k and the principal curvatures are equal and opposite, one sagging and one hogging. The optimal deflected surface in this case is a minimal surface (see A.13) and sections taken at $\pi/4$ to the edge are *straight*. The shape is a shallow hyperbolic paraboloid.

Though the point is not of importance in the present context, a requirement of optimum theory is that any direction which is *not* reinforced by fibres must have a curvature *less* than the constant value of curvature $|k|$ along the fibre direction. This requirement will be violated by the layout proposed in Fig. 5.0 for the cases of $\alpha > \pi/4$. However in any of the examples dealt with later, should this requirement not be met, the effect on the overall optimality can be shown to be very slight.

Thus Fig. 5.0 shows a simple and intuitively sound solution to the problem of the optimal simply supported corner layout, which is truly optimal for $\alpha \leqslant \pi/4$.

Consider next the corresponding layout for the clamped corner.

(b) The clamped edge corner

In this case the only readily obvious feature of the surface shape and spanning layout is that it should have symmetry about the angle bisector.

The clamped edge suggests cantilever spans in the neighbourhood of the edges, with possibly an infilling simply supported span. But what are the proportions and details? These must be found from the requirement that the deflected surface so generated by constant curvature hogging cantilever and sagging simply supported spans, *must* describe a *smooth* deflected surface.

Sketched in Fig. 5.1 is a tentative layout of cantilever beam strips spanning out normally from the edges, and joined by the simply supported span AB which is thought of as similar to the span AB of the simply supported corner. Then DA and EB will have hogging curvature k down into the figure and AB will be a sagging span of constant curvature $-k$ and zero slope point at mid-point C, as demanded by the overall symmetry with respect to the angle bisector.

Next, steps must be taken to ensure that the whole series of adjacent spans, of which DACBE is a typical example, when fitted side by side, do form a smooth surface. The essential condition to examine is the intersection of hogging and sagging spans at the typical point, A. The procedure will be to examine the surface slope vectors for the hogging and sagging sections.

Now DA is a segment of a circular cylinder with zero slope along OD and normal to OD. Hence the slope vector at A for DA is given by the product of k and DA (Fig. 5.2) and is directed along DA.

Consider now CA. The point C is a zero slope point and ACB is a segment of an as yet not fully defined surface. Across ACB there will be some as yet unknown surface slope S, which will be constant if ACB is to be a principal direction.

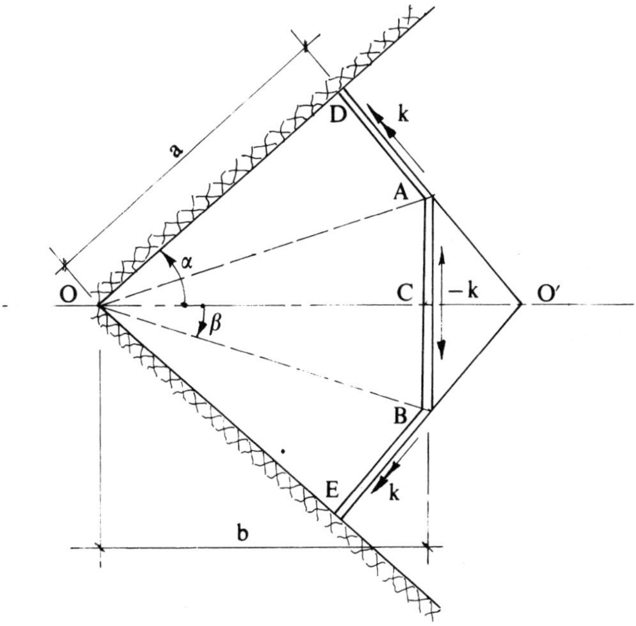

Figure 5.1 Clamped corner arrangement.

The other component of surface slope at A for ACB will be k × AC, since C is a zero slope point, when seen in elevation on ACB.

All this slope information is shown in Fig. 5.2. As surface slope is a *vector* quantity, the vector for portion DA must be identical to the vector for portion ACB, in order that there be surface slope continuity at A. The requirements are that any two selected non-parallel components must be common. Hence

$$bk \tan \beta = ak \tan (\alpha - \beta) \,.\, \cos \alpha \qquad (5.3.1a)$$

and

$$S = ak \tan (\alpha - \beta) \,.\, \sin \alpha. \qquad (5.3.1b)$$

Now for some manipulation and simplification. From the geometry of Fig. 5.1,

$$a \sec (\alpha - \beta) = OA = b \sec \beta$$

or

$$a \cos \beta = b \cos (\alpha - \beta).$$

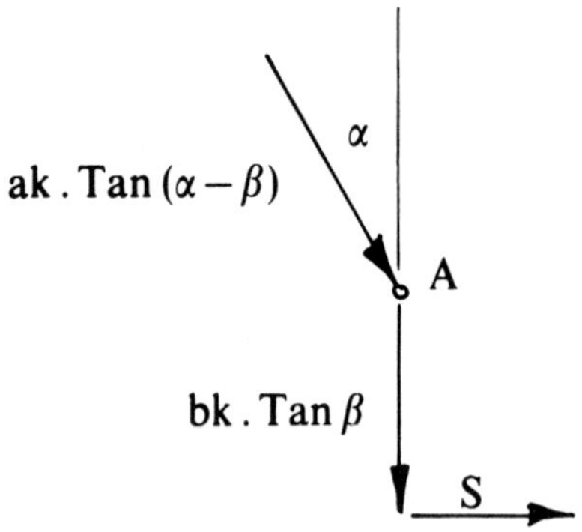

Figure 5.2 Slope conditions at a junction.

Thus (5.3.1a) gives

$$b \sin\beta = b\cos(\alpha-\beta).\tan(\alpha-\beta).\cos\alpha = b\sin(\alpha-\beta)\cos\alpha$$

or

$$\sin\beta = \frac{\sin 2\alpha}{2}.\cos\beta - \cos^2\alpha.\sin\beta.$$

Hence

$$\tan\beta = \frac{\sin 2\alpha}{2(1+\cos^2\alpha)}. \qquad (5.3.2)$$

The expression for β, (5.3.2), therefore fixes the geometry of the corner. For example, if $\alpha = \pi/4$ then $\sin 2\alpha = 1$, $\cos\alpha = 1/\sqrt{2}$ and $\tan\beta = 1/3$.
Exercises: (1) Show that κ_2, the transverse (principal) curvature in the region OAB, is independent of b, is hogging, and given by

$$\kappa_2 = \frac{k\sin^2\alpha}{1+\cos^2\alpha}.$$

(2) Show that the surface transverse slope S in both the simply supported corner and the clamped corner is given by

$$S = k\,(O'\,C).$$

This is a very important geometrical result.

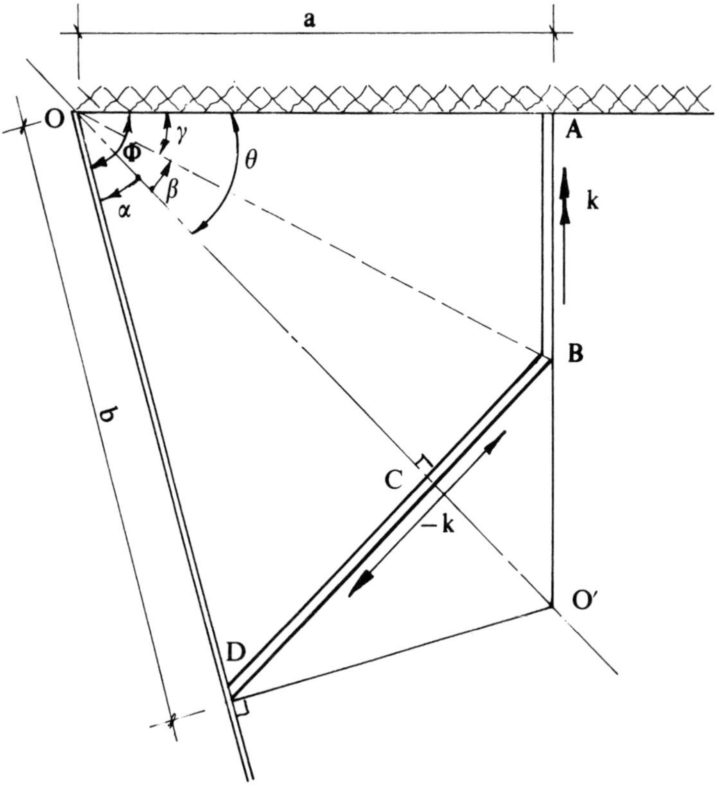

Figure 5.3 Mixed corner arrangement.

(c) The mixed corner

To complete the analysis of elementary corner results, the case of the mixed corner consisting of a simple span meeting a clamped span will now be considered. In this case there is not even symmetry to take as a starting point. Instead, suppose the spanning system is a clamped-edge strip AB, supporting a simple span BC, which in turn is supported on the simply supported edge (Fig. 5.3). This is in the spirit of the previous investigations.

As an extension of these earlier examples the present example can be thought of as a composite of suitable halves from (a) and (b) brought together along their respective lines of symmetry. For these portions to fit together smoothly the transverse curvatures and displacements of the pieces must coincide.

In Fig. 5.3 has been sketched a tentative line of separation. OC, which is half a simply supported corner, semi-angle α combined with a half clamped corner, semi-angle θ. For this to be a smooth junction the angles α and θ must be in the correct relation to give a common transverse curvature.

From the relations developed above this requires that

$$k \tan^2 \alpha = \frac{k \sin^2 \theta}{1+\cos^2 \theta}.$$

Simplifying,

$$\sqrt{2} \cdot \sin \alpha = \sin \theta. \tag{5.3.3}$$

Expressing this in α, Φ terms

$$\sqrt{2} \cdot \sin \alpha = \sin (\Phi - \alpha).$$

Solve for α and obtain

$$\sin \alpha = \frac{\sin \Phi}{(3+2.\sqrt{2}.\cos \Phi)^{1/2}}. \tag{5.3.4}$$

This is the primary relation of this study and allows the whole configuration of Fig. 5.3 to be established.

Exercises: (1) Study Fig. 5.3 further, show that O′D is normal to OD, and that

$$\tan \beta = \frac{\sin 2\theta}{2(1+\cos^2 \theta)}.$$

A very convenient approach is to draw Fig. 5.3 to scale.

(2) For the case $\Phi = \pi/2$ show that

$$\alpha = \gamma, \quad \tan \alpha = 1/\sqrt{2}, \quad \sin \beta = 1/3, \quad b = \sqrt{2}a, \quad AB = BO'.$$

Draw the layout to scale.

The other type of edge which might be found in combination with the previous two is the *free edge*. The guiding rule here is that the constant curvature spans will always be *parallel* to the free edge. How this can be achieved will be seen in examples to follow later.

To conclude, in this section we have discussed the likely spanning directions and deflected shapes in terms of associated constant curvature spans for the pieces of slab in the neighbourhood of corners. These arrangements have not yet been shown to be optimal. Indeed, one of the exercises readers can set themselves is to propose possible alternative spanning arrangements, and explore the geometrical consequences and load carrying capabilities of these alternatives.

5.4 Some complete results

Here the question of proposing possible layouts for the spanning and deflected shape of *complete* slabs will be initiated. Begin with the simplest case, the simply supported circular plate. Geometrically all the requirements for smoothness of the deflected shape built up from constant curvature spans can be met by a shallow spherical surface with sagging curvature $-k$. This can be interpreted in statical terms to mean that any desired system of beams spanning the region is permissible, for example, a parallel system of beams, or if desired a grid of beams. It will be shown later that the strength required, in terms of reinforcement needed is, in theory, unaffected by the choice. Of more interest is the solution appropriate to a clamped edge circular plate. It is relatively easy to show that the only configuration of hogging edge zone with sagging central zone, constant principal curvatures in the fibre directions and smooth deflected shape, is that composed of an outer hogging zone extending to mid radius and a central spherical zone (Fig. 5.4).

A more practical shape is the rectangle. Consider first the simply supported rectangle in Fig. 5.4. There is a combined deflected surface built up from the corner zone discussed in 5.3, with a spherical deflected surface, labelled N (neutral), and a purely cylindrical portion. All these segments imply reinforcement in the direction of the constant curvature, $|k|$, and it can be seen that the resulting spanning system is in effect a one-way grid of beams, except in the neutral zone where a two-way grid is theoretically acceptable (since all directions are principal). But smoothness of the deflected surface is achieved, even if the rectangle is distorted into a diamond shape.

The next stage is to consider the clamped rectangle in Fig. 5.4. Now the layout looks to be significantly more complex but the same fitting together of corner zone, neutral zone and cylindrical portions leads to the layout shown, and this can be demonstrated to possess all the necessary features.

Next in this group of complete solutions can be demonstrated the application of the mixed corner layout developed in the previous section to incorporation into the rectangular slab with alternate edges simply supported, clamped or free. Once again the elements are the same—the corner, the neutral zone and a cylindrical segment. And once again the geometrical properties ensure the smooth fitting together of the pieces.

As typical of the further application of the basic elements of solution via corner geometry to solve other problems, consider the triangular plate in Fig. 5.4. Now the vital role that O′ plays can be best appreciated. This point is common to the three corners, and is the incentre for the triangle. Once again, the essentially

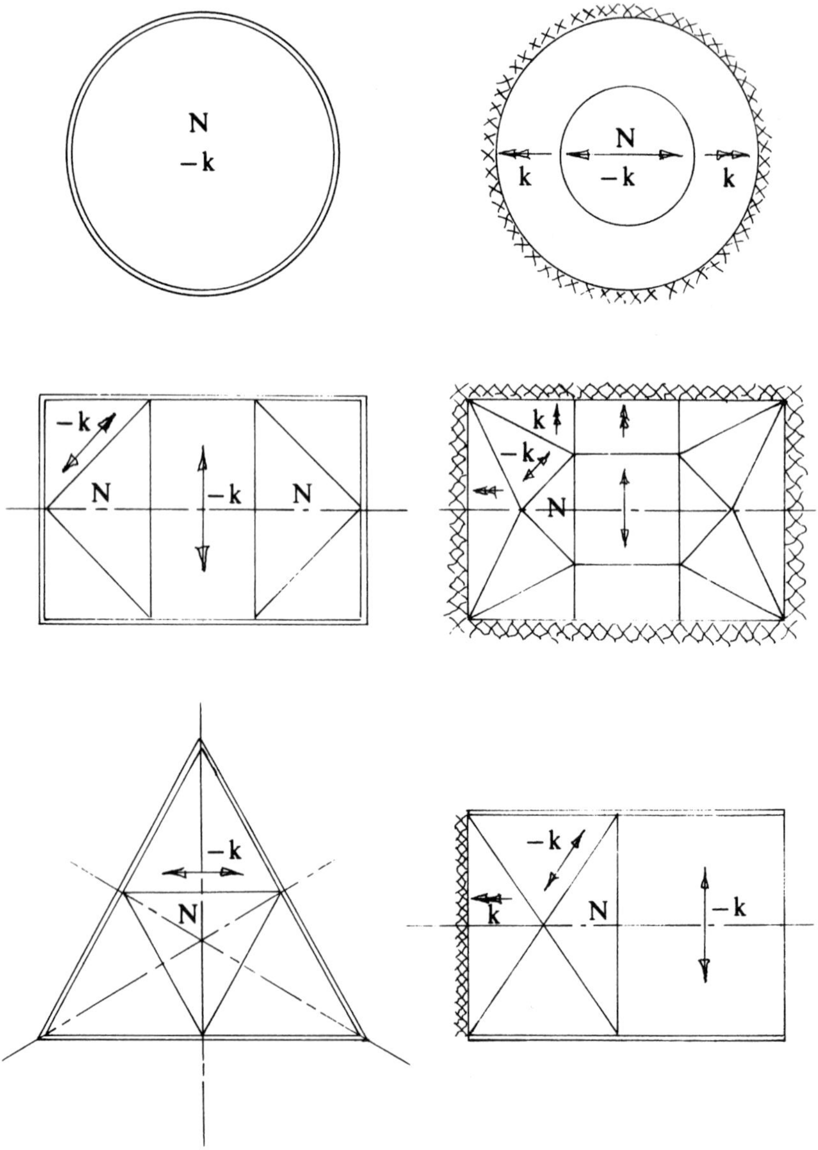

Figure 5.4 A selection of optimal layouts.

elementary geometrical properties are the key to constructing a possible spanning and deflection layout.

5.5 Moment volumes

The measure of optimality used here is the amount of reinforcement needed to support a given load on a slab spanning to the given boundaries. Because attention is to be confined to constant depth slabs, it is convenient to formulate the measure of optimality as the quantity defined earlier, the *moment volume*, rather than the actual *reinforcement volume*.

Moment volume is defined as

$$V_M \equiv \int \left(|M_1| + |M_2|\right) dA$$

where $M_{1,2}$ are the principal moments, and dA is the element of slab (plan) area. This expression is analogous to the expression defined earlier for a beam. If needed, the reinforcement volume (V_R) can be found from

$$V_R = \frac{V_M}{\sigma d},$$

where σ is the common stress in all reinforcement and d is the effective lever arm for all reinforcement, which is assumed to be constant. When the context is clear, the subscript (M or R) will be omitted for simplicity.

As has been emphasized in the construction of feasible spanning arrangements in 5.3 and 5.4, the most prominent feature of the layouts is the local beam-like quality of the problem. The only portion of the layout where this may not be the case is in the neutral zone. Indeed, within the neutral zone the spanning arrangement can be chosen to suit any possible requirements. Elsewhere the spanning arrangement is fixed, since the reinforcement must lie in the direction of the principal curvature, $|k|$.

Consider the moment volume calculation for the simply supported corner (Fig. 5.5). Now, with a uniform loading of p applied, including self weight, the moment volume is given by

$$\begin{aligned} V &\equiv \int |M|\, dA \\ &= 2\int_0^y \int_0^{y\tan\alpha} \frac{p}{2}(y\tan\alpha - x)\,(y\tan\alpha + x)\, dx\,.\,dy \\ &= \frac{p}{6}\tan^3\alpha\,.\,y^4. \end{aligned}$$

When $\alpha = \pi/4$, then $V = pa^4/24$.

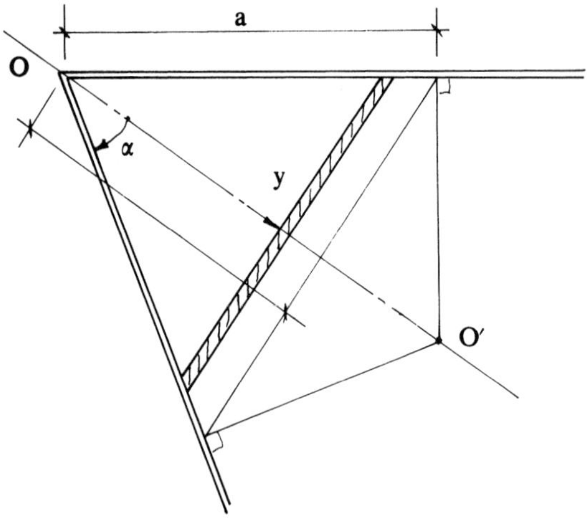

Figure 5.5 Simply supported corner—moment volume.

An alternative formulation is

$$V = \frac{1}{k}\int pw\,dA.$$

Often this formulation is to be preferred because an easy visualization is possible.
Exercise: (1) For the clamped corner including an angle of $\pi/2$ and contained within a square of side a, show that V has the same value, $V = pa^4/24$. (This may seem paradoxical, but recall that this corner supports load over a greater area than the simply supported case. The neutral zone is excluded.)

(2) Repeat the earlier calculation for the simply supported corner, but use now the variable x, Fig. 5.6*a*, and the formulation

$$V = \int |M|\,dA = \frac{1}{k}\cdot\int pw\,dA,$$

to find V. Show that $\Delta = kx(a - x)$ and hence

$$V = \frac{p}{k}\int_0^a \frac{kx}{2}(a-x)^2\,dx = \frac{pa^4}{24}.$$

(3) Repeat the clamped edge V calculation using the variable x, Fig. 5.6*b*, Then

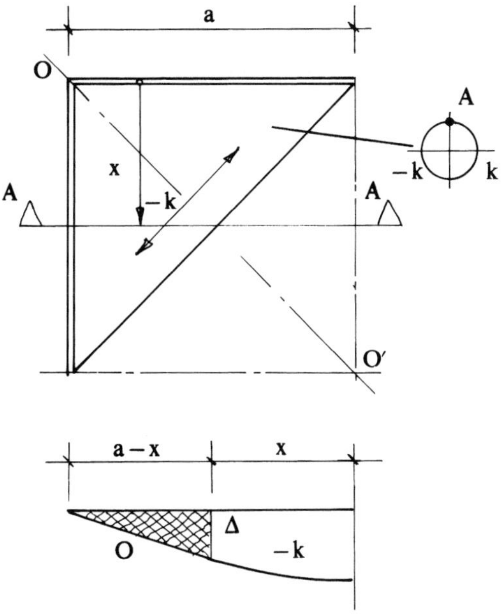

Figure 5.6(*a*) Simply supported corner—alternative section.

$$\text{area}_{\text{Section A}} = \frac{kx^2}{12}.(6a - 5x),$$

$$\text{area}_{\text{Section B}} = \frac{k}{48}(a^3 + 6a^2x + 12ax^2 - 28x^3).$$

Use these results to show that these two moment volumes become 11/768 and 7/256, each multiplied by pL^4, to complete an independent calculation of V.

5.6 Some theory

Thus far the moment volumes computed have not been demonstrated to be minima—the arguments have been intuitive. The aim now is to explore briefly the theoretical reasons why these values are minima.

The moment volume V_M has been defined as

$$v_M \equiv \int (|M_1| + |M_2|)\,dA$$

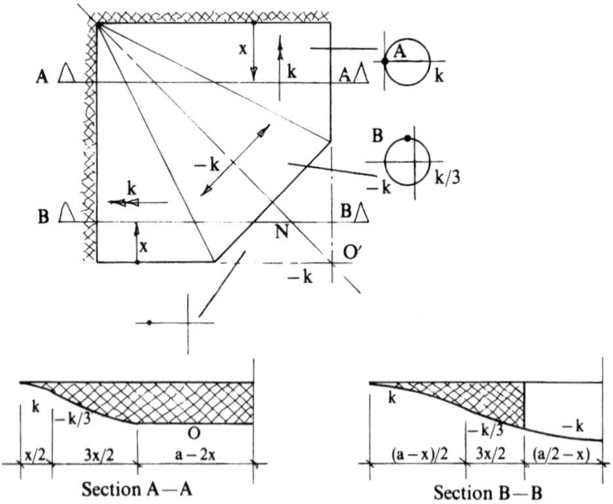

Figure 5.6(*b*) Clamped corner—two sections.

where M_1, M_2 are the principal moments in the plate and dA is the element of plan area. For constant thickness plates and small amounts of reinforcement, such as generally will be the case for reinforced concrete plates, then V_M is related to the volume of reinforcement (V_R) needed by the relation

$$V_M = V_R \cdot \sigma \cdot d$$

where σ is the stress common to all the reinforcement.

The slab is required to support some specified loading, p, which will be assumed to be a uniform pressure, though any reasonable loading can be assumed.

The load will induce bending moments in the slab, $M_{\alpha\beta}$, which will vary from point to point. As the plate deflects, curvatures ($\kappa_{\alpha\beta}$) will develop, associated with a *smooth* transverse displacement field w = w(x, y). Here the expression $(\ )_{\alpha\beta}$ is used to mean the collection of all quantities ( ). Thus for moments, $M_{\alpha\beta}$ is taken to mean M_{xx}, M_{xy}, M_{yy}, with α and β being treated as free indices, and likewise for $\kappa_{\alpha\beta}$. Now, by virtual work, as the load p does work on the displacement w so internal work is absorbed by $M_{\alpha\beta}$ doing appropriate work on the components of $\kappa_{\alpha\beta}$. Symbolically this situation is described by

$$\int pw\, dA = \int M_{\alpha\beta} \cdot \kappa_{\alpha\beta} \cdot dA \tag{5.6.0}$$

where there is a summation on the repeated α and β indices in the right-hand expression. Thus the right-hand side is the sum of terms obtained by letting α be first one coordinate, say x, and β each coordinate (x, y) in turn; and then α is the second coordinate, y, and β again takes on the values x and y. As the result, the moment work term in cartesian coordinates becomes

$$\int (M_{xx} \,.\, \kappa_{xx} + M_{xy} \,.\, \kappa_{xy} + M_{yx} \,.\, \kappa_{yx} + M_{yy} \,.\, \kappa_{yy})\, dA.$$

Each $M_{\alpha\beta}$ and $\kappa_{\alpha\beta}$, being a second order tensor, has principal values and associated directions. Unless something is done about it, the principal directions for $M_{\alpha\beta}$ will not coincide with those for $\kappa_{\alpha\beta}$.

However, suppose M_1, M_2 to be the *principal* moments. Then the moment work term contracts to just two terms

$$\int M_{\alpha\beta} \cdot \kappa_{\alpha\beta} \cdot dA = \int (M_1\kappa_1 + M_1 \cdot \kappa_2)\, dA,$$

where κ_1, κ_2 are merely the components of $\kappa_{\alpha\beta}$ along the $M_{1,2}$ principal directions.

An important step in the argument is the following. If modulus signs are inserted in the right-hand terms then an inequality results, as

$$\int M_{\alpha\beta} \,.\, \kappa_{\alpha\beta} \,.\, dA \leqslant \int (|M_1| \,.\, |\kappa_1| + |M_2| \,.\, |\kappa_2|)\, dA,$$

since it cannot be assumed that M_α, κ_α are always of the same sign. Suppose however, if a choice (c) of M_α, κ_α can be found for which the signs *do* always correspond, that is, are always the same for M_α as for κ_α, then

$$\int (|M_1| \,.\, |\kappa_1| + |M_2| \,.\, |\kappa_2|)_c dA \leqslant \int (|M_1| \,.\, |\kappa_1| + |M_2| \,.\, |\kappa_2|)\, dA. \quad (5.6.1)$$

It should be emphasized that there is no special relation between $M_{\alpha\beta}$ and $\kappa_{\alpha\beta}$. The $M_{\alpha\beta}$ is an equilibrium system of bending moments and the $\kappa_{\alpha\beta}$ is a possible field of curvatures associated with a possible transverse displacement, w.

The essential minimal information derives from an interpretation of (5.6.1). First it is noted that for the minima, the $M_{\alpha\beta}$ and $\kappa_{\alpha\beta}$ must share principal directions and must correspond in sign.

To recover the expression for moment volume the curvatures are restricted to be constant or, should $M_\alpha = 0$, then $|\kappa_\alpha| < k$ allows $|k|$ to be taken out as a factor in (5.6.1) and we obtain.

$$\int (|M_1| + |M_2|)_c \,.\, dA$$

$$\leqslant \int (\text{moment volume for any system not known to correspond})\, dA. \quad (5.6.2)$$

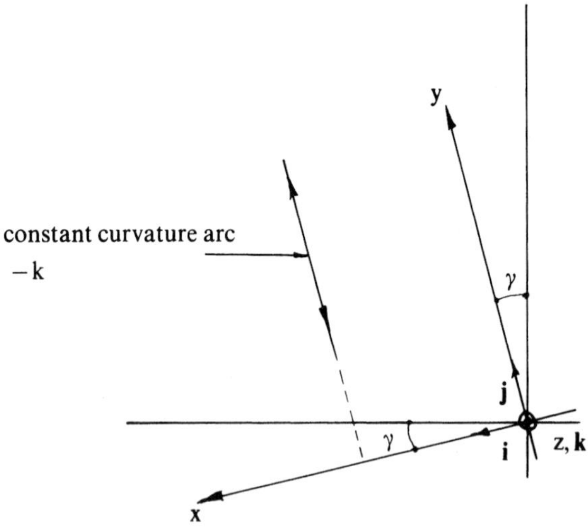

Figure 5.7 General —*k* surface—coordinate axes.

Hence the minimum is attained if associated principal curvature values can be found which correspond with the bending moments but have principal values which do not exceed the constant k value in absolute value.

If the layouts discussed earlier are checked against these criteria, it will be found that the criteria are met and hence the layouts and associated V_M's are minimal.

In addition to these criteria for recognizing an optimal layout there is one important geometrical theorem which will now be proved. This is the theorem relating to the variation of the curvature transverse to a principal constant curvature arc.

Consider the most general context in which such an arc, assumed straight as seen in plan, can occur in an actual surface. The general motion of a constant curvature arc to describe a constant curvature surface can be represented by a radius vector **r** given, see Fig. 5.7, by

$$\mathbf{r} = A(\gamma)\,\mathbf{i} + y\mathbf{j} + \left(a(\gamma) + b(\gamma)\,y - \frac{ky^2}{2} \right)\mathbf{k}. \qquad (5.6.3)$$

Here A; a, b are kinematic degrees of freedom. The coordinates will be taken to be γ, y rather than x, y. Now, for convenience, the notation $d(\ )/d\gamma = (\ )_{,\gamma}$ will be adopted, and the following relations can be shown to hold:

$$\mathbf{i}_{,\gamma} = -\mathbf{j}, \quad \mathbf{j}_{,\gamma} = \mathbf{i}, \quad \mathbf{k}_{,\gamma} = 0$$

and

$$\mathbf{r}_{,\gamma} = (A_{,\gamma} + y)\mathbf{i} - A\,.\,\mathbf{j} + (a_{,\gamma} + b_{,\gamma}\,.\,y)\mathbf{k}$$

$$\mathbf{r}_{,y} = \mathbf{j} + (b - ky)\mathbf{k}.$$

The surface slope is a vector quantity, and hence it is described by *two* components. Remember that these slopes are small, since the deflected surface is shallow. Then, for example, the slope is the y-direction along the −k span is

$$S_\gamma = b - ky.$$

Shallowness means that $kL \ll 1$, where L is a typical span in the problem. Also

$$A = O(L); \quad a = O(kL^2), \quad b = O(kL).$$

The quantities $\mathbf{r},_\gamma$ and $\mathbf{r},_y$ are tangent vectors to the surface along the γ and y increasing directions respectively. They are not mutually orthogonal. However after some manipulation the *unit normal*, **n** can be evaluated to be

$$\mathbf{n} = -S_t\mathbf{i} - S_\gamma\,\mathbf{j} + \mathbf{k} \tag{5.6.4}$$

where $S_t = [A\,.\,b + a,_\gamma + (b,_\gamma - kA)y]/[A,_\gamma + y]$ and $S_\gamma = b - ky$.

The expression for S_t then provides the result that if S_t is the same for two distinct values of y it is the same for *all* values. In this case S_t is given by

$$S_t = b,_\gamma - kA, \quad \text{since} \quad A,_\gamma\,(b,_\gamma - kA) = A\,.\,b + a,_\gamma.$$

When this is so the constant curvature direction is principal. A further ramification of such constant curvature surfaces is that the second principal curvature (κ_2, since $\kappa_1 = -k$ in this case), is given by

$$\kappa_2 = \frac{(b,_{\gamma\gamma} + b)}{A,_\gamma + y} - k. \tag{5.6.5}$$

This result shows that κ_2 is monotonically varying and once again this knowledge proves to be very valuable in studying optimal layouts where the condition $|\kappa_2| < k$ must be satisfied.

5.7 Conclusions

This chapter has been included to indicate to the reader some of the scope of geometry and statics applied to plate problems. In one sense the subject matter is out of place alongside the earlier chapters, because optimal plates have not yet been accepted as practical realities.

However, the reader is urged to regard the content of this chapter as being of a nature of an investment in method, one which may or may not provide dividends. The point, nevertheless is clear that with a minimum of outlay, a method for finding a whole range of closed form solutions to anisotropic plate problems can be built up and used for exploration.

5.8 Exercises

Because of lack of space in the chapter proper a number of results are here suggested to the reader as exercises.

(1) Show that the simply supported square plate (side L) with pressure loading (p) requires a minimum moment volume of $V = 5pL^4/96 = 0.0521\ pL^4$.

Further show that the central (neutral) zone absorbs 40% of this V if reinforced in a diagonal direction across a span of $L/\sqrt{2}$. The corner zones absorb 5% each, and edge reinforcement along the neutral zone boundary the remaining 40%.

(2) For the clamped square plate (side L) with pressure loading (p) show that the minimum moment volume requirement is $V = 13pL^4/768 = 0.0169\ pL^4$.

(3) The simply supported equilateral triangular plate (side L) when pressure loaded requires a V of $\sqrt{3}/256 = 0.00677/pL^4$; the clamped equivalent requires a V of $25\sqrt{3}/18432 = 0.00235\ pL^4$. Establish these results.

(It is suggested in all cases that maximum use be made of the corner requirements worked out, or quoted earlier).

(4) Show that the circular plate, when simply supported and loaded with a uniform pressure (p) requires a V of $\pi pR^4/4$. For the clamped edge circular plate and pressure loading show that $V = 7\pi pR^4/96$.

(5) Perhaps a more rational way of comparing these expressions is to express the quantity pL^4 or pR^4 as pA^2 where A is the plate plan area. When this is done then again the circular plate emerges as the *weakest* plate for given moment volume and either simply supported or clamped edge.

The comparative values are as in Table 5.0.

Table 5.0 Values of moment volume (all $\times\ pA^2$)

Shape	*Simply supported edge*	*Clamped edge*
Circular	0.0796	0.0232
Square	0.0521	0.0169
Equitriangular	0.0361	0.0125

6 Construction and design—a case for new technology

6.0 Introduction

The origins of the preferred construction type

In this chapter we discuss some general *construction and design* principles and then relate these to slab and floor structures. Where specific construction technology is discussed the emphasis will be on new developments rather than established technology.

We begin with a discussion about features that are desirable in building materials and construction methods. These are in the nature of "ideals". But it may not be possible to find materials and methods of construction that can incorporate all these ideals. Then follows a description of what will be referred to as the "preferred" construction type: this is a system compatible with as many of the ideal features as possible and is the new technology referred to in the chapter heading.

Floors made using any system must provide ample stiffness, strength, ductility and inherent passive fire resistance at an affordable cost. All systems should also result in simple components if they are to be competitive and attractive to all the parties: the designers, builders and owners. As we shall see later, the outcome is that our preferred construction type is in important respects different to any existing established construction type. In the later sections of the chapter the preferred construction method is described in sufficient detail to allow typical floors to be designed and constructed using such a system.

Resumé of the earlier chapters

The earlier chapters dealt with the analysis of the internal forces and moments in flat floor-like structures. Such analyses provide input data on forces and especially moments into the design process. Three different methods of analysis have been introduced and discussed. First *elastic* analysis was discussed, providing us with the stiffness (deflection) information needed. The *plastic* analysis was described from which the strength information can be obtained. The discussion began with isotropic plates, which is conventional. This was then extended to orthotropic and anisotropic plates and slabs.

The third method presented was an *optimal* plate theory. In contrast to the elastic and plastic analyses the optimal principles provide an analysis and also a design in the sense that the strength needed at all points in the floor is part of the solution. However the designs obtained by this means are highly orthotropic and are often too extreme for use in routine construction. Despite this feature the optimal solutions do provide useful information about lower limits to the practical reinforcement requirements and specific guidance about reinforcement distribution. The optimal approach is also the only one of the simple models of material behaviour that produces the natural emergence of one-way spanning, unidirectional moment fields such as dominate in practical pre-cast design solutions for example.

Construction thinking post "9/11"

The demonstrated vulnerability of the Twin Towers, especially the floor structures, and the sequence of world changing political and military events that have resulted from their collapse, should be a wake-up call to all professionals whose expertise contribute to the sphere of infrastructure provision. This call we shall interpret to be the need to seek new initiatives for the planning, the design and especially the construction technology of large building structures in the future. It should be a matter of routine that new ideas and methods are being trialled in an ongoing manner. The industries involved are traditionally very conservative, and the timelines faced by parties wishing to introduce new technology are very protracted. In any overall scheme for what is in the public interest, long timelines are not consistent with the proper health of the sector.

The list of "ideal" features discussed below as desirable features of building materials and construction methods are in many cases far removed and unattainable from many present practices in the industries concerned. Here we preview two of these features—ductility and passive fire resistance. Ductility in building structures is measured by the ability of the structure (in our discussion here the floor) to sustain moderate to large deflections before losing any strength. Ductility provision is taken most seriously in countries where the risk of earthquakes is considered high. Yet even in this subgroup the less developed or less wealthy the country is, the less the local building practices achieve a reasonable level of ductility in the new building stock. Steel is itself a ductile material but many steel structures are not ductile in service. The dominance of reinforced concrete in worldwide construction has contributed to many of the buildings constructed being unnecessarily heavy and, in the poorer countries, often resulting in buildings of poor quality. This combination of weight and quality frequently results in inherent vulnerability to damage, and all too often collapse in even modest earthquakes. Many of the largest cities around the world are not perceived to be in earthquake-prone zones and qualities such as ductility have only minor attention paid to them. Our newspapers all too often carry regular reports of modern buildings collapsing.

The collapse of the Twin Towers is however a quite new phenomenon. Many construction and performance related questions are raised by the enormity of the response to that premeditated attack. This is not the place to begin to examine any of these questions. But it seems very evident that fundamentally new construction concepts must be sought to address the situation. Then there is the subgroup of countries where ductility is given very little thought. The reasons for this are probably numerous, but include unwillingness to incur that modest amount of extra initial cost to choose a building system that ensures ductility of a better level than in the past. There is no absolute best choice, but there are superior alternatives available.

Ductility is so very important a feature that we must achieve it in all our construction. And whatever can be said about the importance of ductility, even more can be said of fire resistance of structures. There has been much debate in technical circles for a decade or more around the world about the respective roles for active as against passive protection of life safety in fires in buildings. Some of the debate has been driven by industrial interests wishing to make steel construction more cost competitive compared with reinforced concrete. In recent times active methods of protection, typically water sprinkler systems, have gained in favour. These systems may be the only protection, if the structure itself has little passive resistance.

If the construction type employed does itself have substantial inherent passive fire resistance then the debate could be refocused on the social good, perhaps even necessity, of the need for both passive and effective, failsafe, active protection. Inherent passive fire resistance capacity can be achieved with floors (and other parts of the structure as well) and must not be compromised by the choice of construction type.[1] Where appropriate we shall refer to the "9/11" collapses as a benchmark case where it appears various safety conditions, including inherent fire safety, were not met.

[1] Three words have now been used to describe a flat structural element—*plate, slab* and *floor*. The word *plate* is a commonly used term in theoretical discussion to describe plane usually horizontal, transversely loaded structural elements. A *plate* is near to being a two-dimensional beam. The more specialized term *slab* usually refers to the structural element in a building and usually implies that it is constructed of some type of concrete—reinforced, pre-stressed, pre-cast. *Slab* is almost synonymous with the term *floor*. But *floor* could refer to the wearing surface rather than the structural depth of the floor. If there is a doubt then the term *floor slab* might be used to ensure that the structural depth of the floor is what is being referred to.

6.1 A case for new technology in construction

The present position

Cement, and the concrete made from it, we shall argue are *semi-high-technology* materials. By our definition a semi-high-technology material is one that has further unrealized potential for more efficient and higher grade use. In the case of cement and concrete this is despite more than a century of use of what is essentially the current material. Cement is certainly a strategic material. It is vitally important in all countries of the world and in all economic environments. Cement manufacture is energy intensive and results in massive quantities of CO_2 being released into the atmosphere. The strategic nature of the material and the heavy burden the industry places on the environment should ensure that there is an ever present obligation on all parties, from the manufacturer to the end user, to seek as efficient use as possible of the material.

There are other industries which are also strategic and also place a heavy load on the environment, energy for example. Electrical energy generated by burning fossil fuels is also semi-high-tech in the terms defined here. That industry has many similarities to cement manufacture, on emissions for example—approximately one kg of CO_2 is released into the atmosphere for every kWh of electrical energy generated. A typical domestic consumer's power supplied over a year will account for 10 tonnes or more of CO_2 release to the atmosphere. A difference between the industries is that electricity generators often include with your monthly usage bill the CO_2 attributed, whereas the cement industry is generally secretive about all aspects of production of either the product or by-products: the numbers are commercially sensitive and not available in the public domain. With Mtonnes of cement and TWhs of electricity produced, the CO_2 emissions produced by these two industries is mind-boggling. In a very real sense, for our and future generations health, prosperity and security, we need to ensure that cement properties, as a semi-high-tech material, are further developed to achieve among other outcomes, lower CO_2 emissions, for environmental friendliness and in order positively to discourage wasteful and inappropriate use of the material. The emissions question is returned to later when the preferred reinforced concrete technology is discussed.

Concrete is sometimes used un-reinforced but most applications require it to be reinforced to be fully useful. Steel and reinforced concrete construction have evolved over a period of about a hundred and fifty years. Structural elements such as beams and floors in flexure account for a substantial amount of the concrete made every day. In the average conventional reinforced concrete beam in flexure about half the concrete is not contributing directly to the bending strength since it is in the tension zone and this concrete is likely to be cracked and not contributing to flexure. On several counts this is inefficient: the concrete material is inactive, the tension material must be manufactured and paid for, and the weight of it must be carried as part of the dead load of the structure. Pursuing the above argument about efficient use of strategic materials, a priority aim might be to reduce or even eliminate this portion of the concrete from flexure members in the future. At present there is no priority given to achieving this outcome.

The methods used and the products produced by the reinforced concrete industry are improved from time to time as new knowledge becomes available and finds a place in the construction scene. In general this is a slow process of evolution. This is possibly understandable in an industry in which major changes of the technologies employed pose potential public safety risks during the learning phase of implementation. Also changes may require major expense to re-equip and retrain the workforce. But even if this is the situation it should not stop the search for new and better methods of building and construction. Experience is that progress is often very slow. Industry inertia is an important factor.

Here we are discussing issues relating to new technology for achieving improved building and construction outcomes using concrete reinforced in composite with steel. The materials are traditional but the methods of use in combination may turn out to be substantially different from the conventional use of these materials. In the space available only simple components and uses will be examined. This proposal that there is a need for new technology is being made in the post "9/11" environment, but not primarily because the current technologies are assessed to be technically inadequate. Rather it is because there is accumulating evidence that some of these new products do have superior qualities and offer the prospect of *more efficient use of the concrete material*. This is our primary interest here. What is lacking, and it is a serious lack, are codes of practice that systemize the accumulated experience of their use in service. Introduction of new technology must begin somewhere and the plate-like components in floor and slab systems are candidates for experimenting with new ideas. The components needed are simple and there is scope for improvement of existing floor and slab performance. The decisive role of the floors in the collapse of the Twin Towers is one of several reasons to convince us that it is timely to revisit the subject of floor design and construction.

In the earlier chapters we have been studying the *mechanics* of plate bending. This mechanics provides information about the forces and moments generated in the plate by the applied loads, as well as estimates for the deflection the plate will undergo when loaded in the working range. The elastic calculations are relevant for working load conditions, the plastic calculations relate to ultimate load conditions and the optimum calculations are some guide as to the limits of economy possible. Now we wish to utilize the results of these calculations to make *design* decisions. By *design of the plate or slab* is meant: how to choose the structural thicknesses and other dimensions, and the reinforcement requirements, to support the various load combinations *safely* for a practical plate or slab structure. There are also choices to be made about the construction system to be adopted. The design and construction should be considered in tandem. The broad features of the construction method to be adopted must be known before the design proper can proceed. Always the *safety* of the construction for the required service is the first priority.

New technology—why the need?

Here the choice is made to consider only concrete, and steel for reinforcement, as the construction materials to be employed. These are the most common choices for much construction currently and so there is no novelty here. What will turn out to be new is the manner in which the two materials are combined. In the evolution of reinforced concrete construction, an important early application was in floor slabs. These were initially cast *in situ* construction, meaning that the reinforcement was supported in its final position by some temporary formwork while the concrete was cast in place around it. The formwork was usually not part of the permanent structure and was removed when suitable strength had been attained in the concrete. Formwork has always been an expensive item, and where possible was constructed in such a manner that multiple uses could be achieved. One of our present aims is to eliminate the need for formwork that does not form part of the permanent structure by choice of a suitable alternative type of reinforcement.

The earliest reinforced concrete, that is the material as it was developed and used from about 1870, employed a wide variety of steel reinforcement types including round and deformed bars not too different from the types currently in use. Also used were twisted square bars, expanded metal of various proprietary types and other forms of bar reinforcement. Some of the reinforcement types were protected by patents. There are references given in Chapter 7 where these early reinforcement types are described and illustrated.

In more recent times, pre-cast and pre-stressed concrete components have been extensively developed and employed to speed construction and reduce the amount of temporary works needed. In some countries such components, particularly as used in floor construction, dominate the market. There is an inherent tendency for pre-cast and pre-stressed factory-made units to be large and heavy. Typical units require expert handling while being transported to site and manoeuvred into final position in the structure. Methods for securing the units onto their supports and a number of other practical details need to be carefully considered and attended to if the highest quality of product is to be produced. The connections between units and other parts of the structure are frequently by far the weakest parts of the construction, and this despite efforts over many years to avoid this outcome.

Here we shall not add to the discussion of any of what are currently established design and construction practices. Instead, and in the spirit of much of the mechanics that is discussed in the earlier chapters, we shall describe an alternative method for using these two dominant construction materials, steel reinforcement and concrete, to produce novel forms of structural elements, and floor structures in particular, for possible use in building construction in the future.

The tendency toward one-way spanning systems

The earlier chapters where we deal with plate bending mechanics have included a heavy emphasis on the *two-way spanning action* of the plates such as is implied in an isotropic slab. Much of the discussion has been of plates where there are no preferred directions in the material: as a result it can span in any direction as needed in the particular application. What these studies have shown is the special role that the *shape* of the plate plays in determining the manner in which the plate transfers the loads to the boundary. The other important type studied is the orthotropic plate. These plates do have preferred directions of load transfer, which in the limit may become one-way action. This limit is reached in many of the optimal slabs illustrated in Chapter 5. In practical slabs a one-way spanning action is often the primary means of load transfer. This is especially the case for pre-cast slab floors. There is in general a degree of two-way spanning action in all edge-supported plates. But as has been demonstrated earlier, the shape of the boundary is very important and determines much about the general features displayed by the slab. Another picture emerges when we consider optimal conditions. Now the tendency toward one-way spanning is given a further boost since most of the common shapes of slab, such as rectangular, contain a central dominant zone where optimal theory says the spanning has a preferred direction along which the reinforcement should be provided: hence a local one-way spanning results. This theoretical outcome has not been much appreciated in practical construction. Here we hope to demonstrate that it is consistent with the overall aims of our preferred construction system. In the space available to us here only one-way slab elements will be considered. Extensions to slabs with more complicated spanning features can be made if required later.

6.2 Some "ideals" to be aimed for in construction

We begin with a review of what could be described as "ideal" features for construction materials and methods of construction. Our aim is that the preferred method of construction described later should include many if not all of these features. The relative importance of each feature is not at present being discussed and this list may not be fully comprehensive for all applications. But it is useful to consider what features could be sought, and by what means they could be achieved.

Simplicity

Our list of ideal features begins with the general desire to achieve as much simplicity in construction, both the process and the products, as is consistent with reasonable cost and achieving all necessary performance targets. There are many facets to be considered. Simplicity carries with it an aspect of waste minimization. This is consistent with general aims in an increasingly environmentally conscious world. As responsible professionals we have a role to play in the quest to minimize waste, especially of construction materials, in the whole construction cycle. Use of concrete in construction continues to increase worldwide and implies the ever present need for formwork to contain the new concrete before it sets. This formwork may be temporary (single or multi-use) or permanent. Scope to use permanent formwork for more concrete is a primary aim in the present study.

Automation in the factory and on site

Scope to automate further the range of building and construction processes is at present strictly limited with current technologies. The bar fixing in reinforced concrete construction is a long established construction practice that has little scope for further mechanization. Further construction process automation is a desirable aim and should not be seen as being in conflict with trade and manual skills, and hence with fuelling job losses. Where there is scope to change to a technology better adapted to automated manufacture the opportunity also arises to conserve genuine manual skills for producing higher value products. Construction automation is also linked to the question of achievable tolerances as discussed later.

Concretes of many ages in a single structure

The present norms in much concrete construction result in structures that are composed of concretes of many different ages. Some of these age differences arise from the use of pre-cast components. Others are a consequence of the construction sequences adopted. Essentially concretes of different ages are different materials. Even with the best of techniques the interfaces between two concretes of different ages remain as surfaces, construction joints, where the materials are (intended to be) in intimate contact but are not fully merged. An aim should be to minimize the number of different age concretes in a structure. Pre-cast construction is of a type where concretes of several different ages are routinely encountered. The pre-cast components are factory products; the jointing of components is a site operation. At least two ages of concrete are involved and frequently more than two. Slabs are a favoured application. Most commonly slabs are finished with a topping screed which is structural and comparatively thin, 10s of mm only. The performance of such systems at ultimate load is less well documented than is desirable and some recent investigation casts doubt on the structural performance. The ductility of the general construction type should be further investigated, including load transfer between units and to supports.

The drying out of concrete, cover concrete and strength realization

Associated with the different ages of concrete is the natural tendency for concretes to dry out progressively when exposed to average atmospheric air. Loss of mix water equates to slowing or halting the hydration process and is to be avoided. Yet we are content to allow exposed surfaces, and in particular cover concrete, to dry out, sometimes immediately after setting. This is known to be poor practice and is to be avoided. Taking greater care to avoid drying may be difficult to achieve consistently on site but is worth the effort in added quality of the final product. Controlling or better preventing drying out also has benefits for the shrinkage and creep characteristics of the concrete. A more radical choice might be to eliminate the cover concrete. This sounds extreme and is not an option if the present bar reinforcement is retained. Being vulnerable to damage and, very importantly as the reinforcement protective layer, there remains the necessity to protect it by some new means. There are also implications for fire resistance if the concrete is wet or dry, and also the rate at which carbonation proceeds. Concrete that is not exposed to the air or other carbon dioxide sources will not carbonate. If the concrete is totally enclosed, as we shall be proposing, then free water can be retained to enhance the quality of the concrete and add to fire safety as well as reduce carbonation rates.

Brittleness

Concrete is an inherently brittle material. One measure of brittleness is an inability to sustain tensile strain. For concrete a most important measure of brittleness is the tensile strain limit. This limit is of the order of a few hundred micro strain ($\mu\varepsilon$) for typical concretes. There is also a compressive strain limit but this is around ten times higher. The addition of reinforcement to the concrete is the usual method for achieving a (composite) material with a degree of ductility. Ductility can be recognized as an ability to withstand strain, and more particularly tensile strain: this is the strain resisted by the steel reinforcement when present. Thus ductility is the opposite of brittleness on these measures. If more ductility is required more reinforcement can be added. But there is a further limit—so called "over reinforcement"—after which addition of further reinforcement leads to the switch back to brittleness because the concrete compressive strain limit is then reached and member brittleness returns. Our preferred choice should not be subject to such a limit.

Ductility—under and over reinforcement

Another of our aims is to devise a steel/concrete composite in which the "over reinforcement" limit to avoid brittleness is removed. Then the material will remain ductile for the whole range of steel quantities in relation to the concrete. Ductility comes in various forms and can be eroded or lost in various ways. The Northridge earthquake in California on 17th January 1994 caused widespread damage and a very costly repair bill resulted despite few building collapses and only minor casualties. The more severe Hyogo-ken-Nambu earthquake devastated parts of the heavily populated, modern Japanese city of Kobe exactly a year later, on 17th January 1995. In Kobe there was extensive damage to infrastructure: many buildings collapsed and substantial casualties were sustained. Japan has experienced many damaging earthquakes and much has been learnt by studying their effects on man-made structures. These two recent events have changed perceptions. In the case of Northridge there was generally poor performance of widely used connection-types in new steelwork. This was a blow to confidence and severely affected the use of steelwork designed by the then current practices. In Kobe the many collapses of structures, some quite new, were often due to widespread loss of cover concrete and consequent failure of many lap splices in reinforcement. The cover concrete is traditionally left unprotected. It is itself the protection for the steel reinforcement. But loss of this cover material can initiate much subsequent damage through loss of ductility and then premature failure. One way to avoid such outcomes is to provide other means to introduce the reinforcement needed and to eliminate the cover concrete entirely. This is the option we shall adopt in the proposed construction method about to be discussed. It remains ductile into what for conventional reinforced concrete is the "over-reinforced" range of steel content because though crushing of the compression concrete still occurs at the relevant strains, it is controlled and contained. Hence ductility is retained. This is a radical proposal but radical measures are required if inherent weaknesses in current construction are to be overcome.

Shear, bond and other secondary capacities

Study of a typical textbook on concrete design will show that considerable attention is devoted to investigating the limits on shear, bond and other secondary capacities in reinforced concrete. Unless the rules are followed there is no guarantee that a beam in primary bending will reach its flexural limit before some secondary limit, such as shear capacity. The sort of material we are aiming for is one in which for a flexural member there is every prospect that the flexure limit will be reached before shear or other limits are, and reaching the flexural limit can be confidently expected to be a ductile happening. Reaching the shear limit is usually a typically brittle happening and occurs suddenly with little warning.

Tolerances

Structures are designed and constructed in all sorts of conditions. As a result the quality of the product may be quite poor in some circumstances and places. Inspection and checking procedures, even when well intentioned, may not detect poor quality in the final structure. If not well intentioned then inspection and checking procedures allow considerable scope to deceive. An associated matter is the construction tolerances that can reasonably be expected. If checking procedures can be strengthened and more easily applied, and tolerances more easily achieved, then the quality of the product is likely to be enhanced. The positioning and amount of reinforcement is a typical check that must be made positively but is difficult to implement and the results are uncertain in many practical situations where bar reinforcement is the only choice. Reinforcement that is external is there to be inspected and checked. It is an option with more scope to meet tolerance and quality standards.

Overall depths of floors

The floors in multi-level buildings often support services of various sorts on the underside, and these contribute to vertical depth requirements. An important consideration is the overall depth of the floor, including the structural depth and the services. Underneath the services is likely to be a false ceiling. The total of all these depths should be kept to a minimum. This is often a consideration that is given a high priority and may impinge unfavourably on other features of the floor system. There is often an intense competition for depth available within the total to provide both structure to support the floor loading, and provide for the services, the weight of which is part of the dead load of the floor. Examples of floors such as those designed and built into the Twin Towers of the World Trade Center would provide an especially relevant subject for study under this and other headings in this section of our chapter.

Tension cracked concrete

Earlier we noted the importance of cement as a strategic material and made the observation that there is scope for more effective use of the material. The presence of tension cracked concrete in flexural members is an example of inefficient use of the material and an unnecessary addition to the self weight of the structure. An aim is to minimize the amount of this category of concrete.

Jointing and tubular members

The efficiency of the methods and the quality of the products obtained when jointing members together is of prime importance in achieving acceptable quality and cost targets in construction. The current two dominant technologies, structural steel and reinforced concrete, have evolved their preferred jointing details over many decades of use. In the case of steelwork, the January 1994 Northridge earthquake in California caused extensive and unexpected major damage to new steel structures, especially to the jointing systems employed. The comparative cost of joint details has risen over the years and this has required these details to be simplified to the point that previously unacceptable configurations were being used to meet the competition. If tubular section shapes could be produced at affordable cost for all regular uses then new avenues open for making simple and effective joints. Primarily this is because there is then an *inside* as well as the outside of the members to be exploited in the jointing configurations. Members can slot one inside the other as well as butt against one another, for example. Such scope enables many other jointing options to be developed. Neither dominant technology can at present offer this scope. So any new construction technology should seek to have this option available.

Summary

What we present as our preferred method of construction is the result of seeking to incorporate as many of the ideal features discussed above as feasible. In this search we are led to ask and then seek to answer the following question: given that the structural materials available for us to use are concrete and steel, *what sort of (floor) structures can we devise to incorporate as many as possible of the desired features described above?* Our answer is to require that *the reinforcement be provided as hollow casings inside which the concrete is cast and so is entirely encased within the steel reinforcement in the permanent structure*. Such encasing is not a major feature in any widely used current floor construction system.

By this means all temporary formwork is eliminated, strong and stiff members result and many of the other features can be achieved.

So what is being proposed is a major departure from current practice. To achieve this encasing of the concrete the steel must be supplied in *sheet* rather than *bar* form. Then the *external reinforcement* that results can quite naturally be used as a *permanent formwork* for the concrete. The composite of these two materials combined in this manner has a number of unexpected and impressive features. We shall term this material **Externally Reinforced Concrete** (abbreviated to **ERC**) and spend the remainder of the chapter discussing its characteristics.

6.3 Externally reinforced concrete—the preferred system of reinforcement and construction

The composite material—ERC

We come now to describe the construction material we propose be adopted: what has been termed the "preferred" construction system. The term *external reinforcement* is a description of the distinctive feature of the material, since *the steel reinforcement will be employed in sheet and not bar form* and will be used on the *outside of* rather than embedded in the concrete. The concrete is to be enclosed within the hollow cases into which the reinforcement is to be fabricated. This is a very unconventional choice and may be an unsuitable choice in some applications.

The decision to choose another form of reinforced concrete rather than stay with the conventional bar reinforced product is more or less inevitable once it is seen that several of the key "ideal" features described above cannot be achieved if we limit ourselves to using the conventional material.

In contrast, many of the features that have been listed earlier and described as "ideals" for a construction material can be easily achieved when constructing in externally reinforced concrete. Readers will need to make their own assessments as to whether the product obtained by means of *external* rather than *internal* reinforcement is sufficiently attractive to be used in their constructions. First let us give a brief resumé of the two component materials—the concrete and the steel.

Resumé of concrete properties

The following is a brief outline of the properties of concrete to which we shall refer.

Plain concrete, that is concrete without any reinforcement, is a very brittle material. Even after more than a hundred years of study, the brittleness aspect of concrete is not fully understood. This brittleness, in simplest terms, shows as a marked weakness to resisting tensile strain. The literature seems to favour complicated explanations of concrete properties which follows from the almost universal tendency to explain concrete properties in terms of stress rather than strain. In one dimension there is direct correspondence between stress and strain, but the same is not true in two or three dimensions. Reinforcement is added to overcome this brittleness.

Weakness to tensile strain implies also a weakness to tensile stress. There are however important differences in dependence between strain and stress as measures of brittleness. The typical fracture strain in tension for an average concrete is around 200–300 $\mu\varepsilon$ (micro strain). A commonly quoted value for concretes is a limiting *compressive* strain of about 3000 $\mu\varepsilon$. However, before such compressive strains are reached, tensile cracking will have occurred in directions normal to the principal compressive strain direction. One reason why it is important to distinguish between strain and stress is because for an elastic material, which is an approximation to the working condition for concrete, if all the strains are compressive, then all the stresses will be compressive. But the converse is not true.

This is evident from the small deflection, linear, elastic material equations. These can be written in Cartesian tensor from as $t_{ij} = 2\mu e_{ij} + \lambda\delta_{ij}e_{kk}$, where, t_{ij} is the stress tensor, e_{ij} the strain tensor, δ_{ij} the metric tensor and λ, μ are the Lamé elastic constants. These relations can be inverted to give strain in terms of stress: $t_{ii} = 2\mu e_{ii} + 3\lambda e_{ii}$, when i = j and the summation convention applies. By substitution for e_{ii} obtain $2\mu(3\lambda + 2\mu)e_{ij} = (3\lambda + 2\mu)t_{ij} - \lambda\delta_{ij}t_{kk}$. Then it is evident that if the principal strains are all of one sign then from the first relation, the principal stresses are also all of the same sign. But the negative sign in the inverse relation prevents us drawing a similar conclusion when the roles of principal stress and strain are exchanged. In this sense strain, a purely geometrical quantity, is more fundamental in this context than is stress.

Thus it is quite possible to have tensile strain in an elastic material despite all the stresses being compressive. A simple situation in which this can be observed is a cylinder of the material subject to a greater axial stress than confining stress around the curved surface. A limiting case is when the stress on the curved surface is zero, i.e. stress free, as in the so-called unconfined compressive testing of a material such as concrete. Then the strains in all directions normal to the applied axial compressive stress are tensile, despite there being no tensile stresses present. Long before the maximum load in compression is reached, tensile cracking will have occurred on planes containing the direction of the compression force on the cylinder. The standard measure of concrete strength is the cylinder (or sometimes the cube) strength. But what is really being observed is progressive tensile collapse as the tensile strain cracking accumulates until such a stage is reached that a mechanism of collapse can form.

The role of geometry, through strain, thus emerges, once again, as very significant. The key role of geometry has already been noted in earlier chapters, in relation to the lower bound calculation of collapse loads and in the study of optimal plates.

In use most concretes are (tensile) reinforced to overcome the brittleness. This can be achieved in a number of ways. The most common method is by means of embedded bar reinforcement, though this is not the type of reinforcement we plan to adopt here. When reinforced the (now) composite steel/concrete material displays a measure of ductility. The property of ductility, as we have discussed elsewhere, is the ability to withstand strain, and particularly tensile strain. It is a most desirable feature for the material to possess.

Concretes have densities about 2.4 times the density of water and compressive strengths in the 10s of MPa range, up to possibly 100 MPa or more, in special cases. Typical Young's moduli for concrete are about three orders of magnitude more than the compressive strength. A figure we shall commonly use is $E_c = 25$ GPa. Various empirical formulate relate Young's Modulus to the compressive strength by relations of the form, $E = \text{const}\sqrt{f'_c}$, or similar, f'_c being compressive strength.

When the concrete is mixed and cast into whatever formwork is being used, water is one of the constituents—the so called mix water. In theory this must not be more than is necessary to achieve full hydration of the cement, and this can be calculated. Usually, in practice, this theoretical, ideal amount of water is too little to produce a mix capable of being compacted in the formwork. The concrete is said to be insufficiently workable. Additional water is therefore added but this should be kept to a minimum since the final strength and quality of the concrete reduces as the water amount increases beyond the theoretical amount for full hydration. Also, excess water leads to voids in the concrete as the excess migrates away and eventually evaporates. What we shall see for the use to which we plan to put the concrete is that any excess water will take a great deal longer time to evaporate, and while present in the concrete is a safety feature for the material in the fire situation. Any water, including water bound in the hydration products, is potentially available to be boiled off in a fire. This contributes to a heat sink to help combat the effects of the fire.

Water loss and drying out, from what ever cause, is accompanied by shrinkage of the concrete mass. Although any shrinkage, expressed as a volumetric strain, is apparently very small it is significant and should be avoided if possible. Also concrete under stress tends to creep, which is a time-dependent movement. Again the strains are usually small but are significant and again should be avoided if possible.

Concrete begins to set quite soon after casting and gains strength rapidly and then progressively over about the first 30 days after casting. This 30-day figure is an industry agreed figure—the strength development continues for longer, possibly for years. The formwork containing the concrete, if not permanent, is removed at the latest 30 days after casting, and in many cases much sooner. It is recommended that the concrete, termed "green" at a young age, be kept wet for as long as possible. This is to ensure that the outer few mm, which is part of the concrete cover to the reinforcement, does not dry out and hence cease hydrating. When hydration ceases so too does strength development. The process is not reversible, and later wetting will not reactivate the strength development, at least not as vigorously or completely as initially.

This natural tendency to dry when exposed to the atmosphere is one of the reasons for choosing to employ all-enveloping permanent formwork of the ERC type in the construction. Exposure to the atmosphere also has other detrimental effects which are also avoided.

Resumé of steel properties

The other material we shall specify is steel. Modern steels have various compositions and properties. Changes to the carbon content and addition of alloying elements have a major influence on the properties of the resulting steels. We shall be concerned entirely with low to medium strength steels, so called mild steels.

Mild steels are inherently ductile in tension and it is this quality that is shared with the concrete when the steel is used as reinforcement. Unlike the majority of reinforcement used in current practice, which is used in bar form, here we shall use only sheet steel. This thin steel, from gauge thicknesses up to perhaps 4 mm thickness, is supplied by the manufacturers as a coiled product. In use, the coil of steel is unrolled and straightened then fabricated into the required shape. The coil has many advantages over the cut sheet as the source material, including the freedom to obtain whatever length the application requires rather than be limited by some maximum available sheet length.

The typical strength of mild steels in tension is from around 200 to 500 MPa. The Young's modulus is typically about 200 GPa, again about three orders of magnitude greater than the strength. The density of steel is about 7.8 times that of water, so about 78 kN/m^3 based on a gravitational constant g value of 10 m/s^2.

Specific features of externally reinforced concrete

The typical practical situation we consider is that the reinforcement will be in the form of thin steel sheet with thickness in the range 2–4 mm. Such sheet is usually available from the manufacturers as coiled material as well as cut flat sheet, and in various widths. Uncoiling and working with the long lengths when needed is a well-established industry practice. Being thin the steel is readily cut and folded into the shapes required. For most purposes steel of this thickness range can be handled as sheet metal and this implies a relative ease of handling compared with thicker material. In the longer term, as experience is gained, and depending on the number of similar components required, roll-forming and eventually full mechanization of fabrication become possible. Bar reinforcement in conventional reinforced concrete can commonly range in diameter up and beyond 30 mm, though in floor slabs the bar reinforcement is normally of much smaller diameter. The bar is generally used in straight lengths and these are of a maximum of a few meters long. If longer bar lengths are required the usual procedure is to lap splice or sometimes weld two lengths to produce longer lengths. Lap splices are usually created on site in the final position, whereas welding may be a shop operation. Sheet sourced from a coil avoids the need to lap the material. This is an important simplifying and attractive feature for external reinforcement technology.

The general procedure is that the sheet steel will be folded into closed rectangular tubular shapes. There are several methods of approach available. The only one we shall discuss is to form these box shapes as pairs of open "U" shapes which are then mated as pairs and MIG welded along two longitudinal, opposite edges to form a rectangular hollow section with double flanges and single thickness webs. A typical section is shown in Figure 6.0. What can be achieved will depend upon the equipment available. There are limits on the shapes that can be folded which are well known in the industry.

The primary use we put the steel to is as a thin sheet in tension. Also it is at maximum eccentricity since it is on the outer face of the member. The bar steel used as conventional reinforcement is primarily in tension but not at maximum eccentricity since it is buried in the concrete. In conventional steel construction

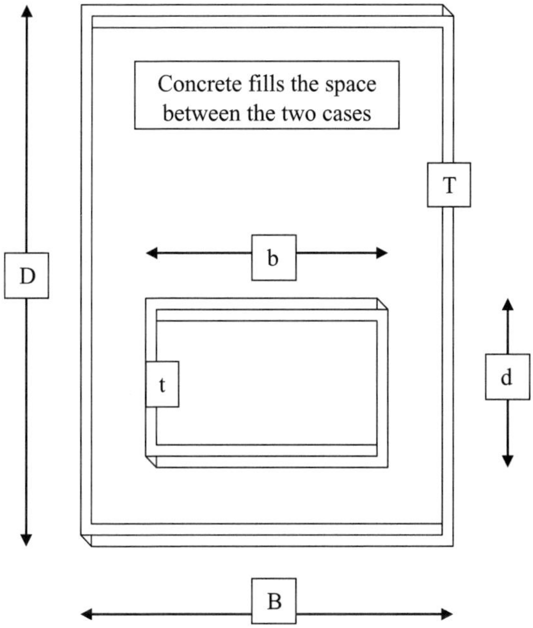

Figure 6.0 Externally reinforced beam section.

care must be taken to ensure that any thin steel section working in compression does not buckle. Some of the ERC steel case is in compression but this steel will not be so governed by buckling because it is always associated with concrete as the main compression component.

This approach to reinforcing concrete can be applied to all forms of reinforced concrete construction although some applications are better adapted than others: a selection of references describing various uses is given in the Bibliography, Chapter 7. Here we are considering only slab-type elements for use in floor construction. Most experience reported in the general literature of concrete-filled tubes has been in using these types of element as compression members. In the external reinforcement literature the primary use of the concept has been as flexural members as is being considered here. The simplifications able to be achieved with these flexural elements can only be appreciated with experience of them. The strength, stiffness, ductility and potential passive fire resistance of the product are especially impressive.

Typically such box shapes for floors will be about as wide as they are deep: see again Figure 6.0. In the simplest construction, units would be laid, side by side, and spanning across the width between walls or supports to create the floor. Such an arrangement would create a one way spanning slab. If required, two-way action could be achieved by connecting the side-by-side units one to the next. This can be done in a variety of ways including transverse pre- or post-stressing. Once in place, the empty box sections are then filled with concrete to give the finished structural elements for the floor. As part of the design brief, the empty hollow cases can be made to be self-supporting while they are filled with the concrete. Costly temporary support systems are then not required.

Such an arrangement has some of the features of a conventional prefabricated or pre-cast system. One of the important features of the system described above for creating externally reinforced concrete members is that when handled into final position on site the steel box sections are normally empty. The steel boxes are relatively light and typically account for about 20% of the finished member weight, 80% being the concrete filling the case. These shop fabricated, light, empty, prefabricated, steel tubular members as transported to the site and handled into position give a cost saving and greatly contribute to site safety since only light components are being lifted and manoeuvred. The major weight in the finished structure resides in the concrete and this is pumped in to complete the construction after assembly of the external reinforcement, which is the permanent formwork for the concrete. Depending on spans, thicknesses and depths used, the empty box sections may require temporary support until the concrete has achieved a sufficient strength.

The further development of the system is to use an inner, smaller casing to increase the steel area available as tension reinforcement while at the same time displacing tensile-zone concrete. This inner case, which we shall refer to as a duct, remains empty of concrete and is available for the secure housing of services. It is also a key element in building passive fire resistance into the section. Figure 6.0 shows a typical combination of outer case and duct. In a fire situation the duct could be flooded with water to serve both as a water source and as a heat sink.

External reinforcement and cement properties

Earlier reference has been made to the demands the manufacture of cement places on the environment. Primarily this derives from the production of CO_2 as a bulk by-product. The manner in which the concrete is employed in externally reinforced components, where the concrete is not exposed to the atmosphere and the steel reinforcement is not embedded in the concrete, therefore offers the potential to produce a less alkaline cement that could continue to provide a satisfactory product in ERC applications.

In ERC applications, because the concrete is not exposed to atmospheric air the rate of carbonation is vastly reduced. Carbonation in current concretes is undesirable because it is associated with a reduction in the alkalinity of the cover concrete, and this in time produces conditions for reinforcement corrosion.

The concrete contents of an ERC member, due to the associated alkalinity, provide an important amount of corrosion protection to the inner surface of the steel casing. Somewhat less alkalinity could also suffice.

So on two counts at least ERC-type technology could provide a window of scope for the use of less alkaline cements. This is turn implies less carbon dioxide production during manufacture of the cement. There are associated favourable energy balances for the processes.

6.4 Section design for externally reinforced concrete members

Preliminaries

The concrete-filled rectangular steel section proposed here for use in a typical floor system is to be designed to resist bending as the primary action. Experimental observation shows that such members are remarkably stiff and strong, and are very ductile. All of these structural qualities are needed if the material is to be competitive in practical use. The form of construction has many other virtues that can translate into cost savings. Discussion of these issues is given in the references included in Chapter 7.

For practical purposes, despite there being no special shear or bond connection provided between the outer steel case and the concrete filling, the assumption made in the following analysis is that there is no slip between the case and the concrete. The action of filling the empty casing with concrete, and ensuring that there are no pockets of trapped air or other voids, is a skill that must be acquired and is a sufficient insurance that there will be acceptable qualities of adhesion between the casing and the concrete. Using this mechanical model for section behaviour a transformed section approach is appropriate to evaluate the stiffness properties of the section.

ERC section stiffness calculation—deflections

A typical ERC section of the simplest type is shown in Fig. 6.0. We wish to write down an expression for the stiffness of the section. Stiffness is the product EI for the whole section, where E is Young's modulus and I is the second moment of area. Because we are dealing with a composite there are at least two Young's moduli and second moments of area to be considered. The EI we calculate will contain a contribution from each material. Initially the concrete will be uncracked, and the steel though very thin will not be affected by buckling under compressive stresses. This stiffness will be termed the *ideal* stiffness. The neutral axis will pass through the combined centroid, at the mid-height for a doubly symmetrical section. This is the simplest stiffness to find for sections of the type we propose to use. The stiffness of the uncracked section is the maximum possible, but is of the least practical usefulness since soon after any bending moment is applied some of the concrete will soon crack.

The moment to cause the concrete to crack we shall denote by M_{cr}. It is usually quite small compared to the ultimate moment which we shall denote by M_R, the moment of resistance. So for most applied moments the concrete will be cracked and once cracked the calculation for stiffness must first find where the neutral axis is located across the section. There is a two-stage process as the moment increases. After the cracking moment has been exceeded the moment can increase with increasing strain across the section until the steel in compression reaches a critical buckling value. As the moment increases further, more steel buckles in compression until finally the moment of resistance is reached and the beam has reached the maximum resistance it can offer. Throughout this loading cycle the beam stiffness reduces as first the tension concrete and then the compression steel reaches a limit. In this section we calculate the stiffness relating to these three stages as the moment increases.

The typical section of a floor element is shown in Figure 6.0. The first calculation is to find the neutral axis depth. For this the expression for force equilibrium is required. If the extreme fibre strain in the compression steel is denoted by ε_s, then the distribution of strain and hence stress across the section can be written down. The strain distribution is assumed to be linear as is usual, and the stresses will be assumed to be elastically related to the respective Young's moduli for this calculation.

Force equilibrium then becomes:

$$\frac{(B-2t)(h-T)E_c\epsilon_{s.}(h-t)}{2h} + \frac{Bh^2E_s.\epsilon_s.-(B-2t)(h-T)E_s\epsilon_s(h-T)}{2h}\theta =$$
$$\frac{(B-2t)(D-h-T)^2(\Omega E_C-E_s)\epsilon_s}{2h} + \frac{B(D-h)^2E_s\epsilon_s}{2h} \qquad (6.4.0)$$

Terms on the left are compressions, on the right tensions, first concrete then steel. For simplicity the duct has been omitted from this equation.

The expression (6.4.0) is a quadratic in the unknown, h, the neutral axis depth. Solving for h we can then find EI. A spreadsheet is a suitable arrangement to organize the calculation. In the above expressions, the factors Ω and θ take the value 1 to begin with. When the concrete has cracked in tension then $\Omega = 0$. As the loading continues to increase, the steel in compression will finally buckle, then $\theta = 0$.

In some applications there is good reasons to construct the section as two nested (i.e. one inside the other) tubular shapes. Typically if the overall dimension of the section is 150 mm square then the inner tube might be 50 mm square. This could be positioned concentrically or in the tension zone and the annulus between the inner and outer tube is filled with concrete. Such a shape has a number of useful features. First the amount of tension zone concrete is reduced. This reduces dead weight and cost of concrete without affecting strength. Also, the inner duct in the tension zone adds to the section strength. The inner duct is expected to be lighter than the displaced tensile concrete and so there is an overall weight reduction. Another and important benefit is that this inner duct is well protected from fire and can operate with the concrete to provide useful passive fire resistance for the section as a whole.

Example: Consider a section 150 by 150 mm, folded from 2 mm thickness coiled steel sheet to give 4 mm flanges, see Fig. 6.0. Inside this tube let there be a concentric 50 by 50 mm inner tubular duct constructed in the same manner and also from 2 mm sheet. With Young's modulus for steel of 200 GPa and the concrete of 25 GPa, the starting value of h can be seen by inspection to be 75 mm. The $(EI)_0$, the uncracked value of the stiffness then computes as 2375 kNm^2. For the cracked section h reduces to 61.9 mm and the corresponding $(EI)_1$ reduces to 1833 kNm^2. Finally after the compression steel buckles, the h increases to 77.8 mm and $(EI)_2$ reduces still further to 1213 kNm^2. If there is an inner duct, as will frequently be desirable, some advantage can be obtained by moving it down in the section, deeper into the tensile zone. If the duct is omitted the various values become: $h_0 = 75$ mm, $(EI)_0 = 2341$ kNm^2; $h_1 = 59.3$ mm, $(EI)_1 = 1763$ kNm^2; and $h_2 = 78.6$ mm, $(EI)_2 = 1165$ kNm^2.

Once the section stiffness is known then deflection of the structural element can be calculated by the usual formulae. Typical service life deflections should be kept below about span/250. The effects of creep and shrinkage of the concrete should be incorporated. This can be allowed for by adjusting the Young's modulus of the concrete.

ERC section strength calculation—the moment of resistance

The strength of ERC sections can be evaluated by adopting a mechanical model that is closely allied to the well-established ultimate behaviour model for conventional reinforced concrete sections. The difference is of course that here we are working with a sheet steel casing, or perhaps two, rather than bar reinforcement. But the concrete behaves in very much the same way as in a conventionally reinforced member. Thus we shall seek to locate the neutral axis such that all steel in tension yields, steel in compression yields and may eventually buckle, while concrete is only active in compression and then under a uniform stress, the cylinder strength, f'_c, over 0.85 of the depth of the compression zone, measured from the outer concrete compression surface, Fig. 6.0.

We prefer to err on the conservative side and so will exclude all compression steel from contributing to the moment of resistance. Now use the notation shown on Fig. 6.0 to write down the following expressions for forces and moments across the section.

$$x/0.85 + y + T = D$$

and

$$f'_c x(B - 2t) = f_Y((B - 2t)T + 2yt)$$

from which we can solve for x and y. For a simple beam consisting of an outer case only, no internal duct and concrete filled the moment of resistance is given by

$$M_R = f_Y[T(B - 2t)(D - 1.5T - 0.5x) + 2ty(D - T - 0.5(x + y))]. \quad (6.4.1)$$

Example: Consider again the 150 by 150 mm section studied for stiffness properties in the ERC section stiffness calculation, above, and shown in section in Fig. 6.0, but without the internal duct. If the steel of the casing has a yield stress of 350 MPa and the concrete filling a cylinder strength of 30 MPa, then the moment of resistance, M_R, for this section, based on the above equations evaluates at 29.8 kNm. Here x = 67.8 mm and y = 66.2 mm. A duct can be included and will add somewhat to the moment of resistance.

As the applied moment increases towards the ultimate, maximum value, first concrete cracking will occur then yielding of the steel will begin and this will progress until all the steel in tension has yielded. The traditional "over reinforcement" restriction to avoid reversion to brittle behaviour of a reinforced concrete member is not a limitation here. Because the steel sheet of the casing is comparatively thin, there will be a tendency for the steel to buckle in compression as the moment of resistance is reached. The behaviour is predictable and the approach toward the ultimate moment is associated with increasing deflection and strain. These are the characteristics of a ductile material. The conservative assumption will usually be made that compression steel makes no contribution to the ultimate limit state value for the ultimate moment strength, M_R, the moment of resistance. An exception might be made if the section of the duct is small, say 100 mm square, when the experimental observation is that such sections can reliably sustain compression stresses approaching yield and do not buckle. But this is a minor contribution to the section properties and the no compression steel conservative assumption can still be applied.

Loadings and a check on the adequacy of the design for a 6 m span

The structure resists two types of loads—the structure's self (or dead) weight and the imposed service (or live) loads. These latter are the occupants and their associated furniture, equipment etc., and the loads imposed by nature such as wind, snow, earthquake, dynamic from equipment or similar effects, that the structure must resist. It is usually expected that dead loads are better defined than live loads and so any factors of safety or load factors used to build in a margin of safety are less for dead than for live loads.

Dead loads will be denoted by G and live loads by Q. Then a typical requirement might be that the design load to be resisted will be 1.2G + 1.6Q, where the 1.2 and 1.6 are load factors to ensure some margin of safety. Recall that the density of steel is about 7.8 greater than water, and of concrete about 2.4 that of water, so 78 and 24 kN/m^3 respectively. The 150 by 150 mm steel outer case weighs 0.138 kN/m, the concrete filling 0.498 kN/m and hence G = 0.138 + 0.498 = 0.636 kN/m. With a live load on the floor of 2.5 kPa, on the 150 mm wide section Q = 0.15 × 2.5 = 0.375 kN/m. Hence loading = 1.2G + 1.6Q = 1.36 kN/m. On a 6 m span the simple moment maximum is $1.36 \times (6)^2/4 =$ 12.24 kNm. The moment capacity is 29.8 kNm, so ample strength available. Next calculate the deflection at mid span of $5 \times 1.36 \times (6)^4/384(EI)_1$: this comes to 13.0 mm. A safe working limit is about span/250 or 6000/250 = 24 mm. The 13 mm will grow with time, due to creep in particular, so this looks like a suitable design for a 6 m spanning, one-way slab.

Strengthening, demolition and recycling of ERC components

Reuse of existing buildings for other purposes may require strengthening of parts of the structure. In addition to all the known methods for strengthening existing structures, ERC as a rather different material to conventional steel/concrete composites offers new ways to effect some strengthening tasks. Cladding an existing column or beam with an outer case and then filling any void with a suitable concrete or mortar mix is one type of strengthening that can be relatively easily achieved. When this is attempted the benefits of lightness of most steel casings and the relative ease with which these components can be handled and fixed on site is a further example of the general philosophy behind this ERC composite.

The process of building using most materials generates waste. If an alternative system offers scope to cause less waste in the building process then there should be a cost benefit to be obtained. Demolition of an existing structure is often an even greater source of building waste. Such wastes are an environmental burden and represent a substantial portion of all wastes that communities have to deal with. Conventional reinforced concrete structures are a particular source of waste in quantities and of a type that is difficult to deal with effectively if the aim is to recycle the materials. An associated feature is the "social pollution", such as the noise, that demolition often generates. When the need to demolish arises there is greater scope to recover the materials from an ERC-type structure, and with much less associated noise pollution, than there is from a conventional reinforced equivalent. Also the form in which the materials are recovered is more environmentally friendly and the recovered material is capable of being of a superior quality for subsequent recycling.

The dust generated when conventional reinforced concrete is broken up is a hazard and is greater than it need be. Some of this dust may arise because there is a significant content of un-hydrated cement in the concrete. Common practice over decades has resulted in many concrete surfaces drying prematurely, and as a result some of the cement content remains un-hydrated. This is wasteful of cement and leads to unnecessarily high dust contents in the broken material at demolition. For ERC on the other hand, because of the enclosure of the concrete, the mix water is available for much longer to enable hydration to proceed than is the case for conventional reinforced concrete. More complete hydration will result in less cement dust in the debris when ERC concrete is eventually broken up, and there are other benefits.

A similar situation is true of the reinforcement. Bar reinforcement can only be recovered if extensive breaking of the concrete is undertaken. For the more heavily reinforced components recovery may be near impossible at an economic cost. The ERC component on the other hand can be very readily separated into the concrete and steel fractions by simple cutting or slicing of the cases since the concrete is not bonded onto the steel.

Fire resistance of ERC

A topic of considerable public interest in relation to buildings and infrastructure, especially where there is public access, is the question of fire resistance. In the last quarter of the twentieth century there were changes made to the fire regulations in many countries with the aim of enabling active systems such as water sprinkler systems to be employed, possibly as the sole means, to protect multi-storey office building and their occupants in the event of fire. Passive protection especially of structural steel frame buildings by the use of spray-on types of fire protection, intumescent paint systems and especially concrete encasing had been the norm but these means of protection have been less favoured in the recent past because of their unfavourable cost environment.

The tragic sequence of events on the 11th of September 2001 in New York resulting in total collapse, primarily due to fire damage, of the Twin Towers is a case study we should all be familiar with. One outcome probably should be a speedy return to passive protection. But it should not be of the previous spray-on type as used in the Twin Towers. As was seen to be the outcome on that fateful day, spray-on passive protection is itself very vulnerable to damage or loss. A quite different scenario is offered by ERC in terms of inherent passive fire resistance, meaning resistance that is an integral part of the structural unit and not an add-on that could be omitted, damaged or lost, depending on the circumstances of the fire emergency. See Chapter 7 for references.

The option is available for every ERC member to be assembled from two or more cases, nested one inside the other with concrete filling the annulus between the outer and inner cases. In the fire situation, parts of the outer case will heat up as will any unprotected steel. As it heats the steel softens and loses strength, but does not melt. Generally the minimum concrete thickness in any ERC member with a duct adjacent to the outer case will be 50 mm or more. Such a thickness is sufficient to ensure that the steel of such an inner case or duct embedded at least this depth will not be affected by fire even in an extended duration of hours. For purposes of fire resistance one design scenario could be to regard those parts of the outer case exposed to the fire as sacrificial and so too the outer approximately 40 mm of concrete in contact with this hot steel. These parts of the section would lose strength in an extended fire. All the materials interior to these outer portions will then be essentially unaffected by fire and can be designed to resist whatever actions the codes demand as needed to resist dead and live loads in the fire situation.

Because the steel used in the cases is thin and only limited parts of the case will be exposed to the fire, the transmission of heat by conduction through the steel, despite it being a good conductor of heat, is relatively limited. This feature can be explored by calculation but is most forcefully demonstrated by actual heating of test pieces. For typical sections and steel thicknesses, and heated so as to cause the exposed steel to glow red as in a fire, it is probable that the remainder of the case not being heated will be sufficiently cool to be touched by hand only a few 100 mm from the edge of the red-hot zone even after prolonged exposure to the fire temperatures. This knowledge gives confidence that even if the outer case of a typical ERC member is not fire protected, the residual section of the whole, less the heat-affected zone, can function as a load bearing component in the most severe of fires. There is also the option to fire protect the vulnerable parts of the section with any of the means already available to protect steel components passively. The best case scenario from the safety viewpoint is to provide both passive and active protection in practical cases.

Other features

The ERC material has scope for much wider use than just in floor slabs. General framed structures can be built using this material. Experimental frames have been constructed and tested. The results of these investigations are reported in references in the Bibliography, Chapter 7. A subgroup of framed structures is domestic building. In many countries a brick exterior with timber framed interior construction is the local norm in house building. The all-timber house with both framing and cladding in timber is the tradition in some cities and countries. A construction system using ERC could be used to provide both an economic alternative while offering greatly enhanced fire resistance than some of the current house building methods can achieve.

ERC has the scope to meet the same levels of strength as other steel-frame housing but can offer markedly superior fire resistance because of the heat sink properties of the concrete filling the cases. In addition to the framing for the house, cladding can also be produced ERC-fashion from steel folded into suitable shapes not too different from the timber sizes presently used, to allow a concrete filling. There are benefits this can provide both as insulation from the normal external environment and fire resistance in the extreme event. There is also the scope to factory assemble all the framing and cladding as a unit but empty of concrete, transporting these assemblies to site, positioning them and then finally filling with the concrete.

During the lifespan of an ERC structure corrosion protection will be necessary, as is the situation with any steel exposed to the natural environment. Currently available systems applicable to steel are applicable to ERC. Only the outer surface needs protection. The inside is in an inherently protected alkaline environment because of the contact with the concrete.

There are several other features of ERC that could be discussed. We shall just mention some of them and maybe these brief mentions will attract readers to examine them and develop them in greater detail. These are further illustrations of the versatility of the material. A first point to make is that we have taken for granted that the filler inside the external reinforcement is a regular concrete. If there is a need to conserve cement, or if none is available, or there may be other reasons, then a filler that has reduced amount or even no cementing agent can be used and many of the features of "conventional" ERC remain, though in lesser amount. This is true for members of all types, including flexural members. In some low-technology applications, or in temporary structures, un-bonded filler may serve a useful purpose.

Another feature we have not given much discussion of is the potential to vary the section properties along the length of a member. There is considerable scope to lay additional flange plates in the empty cases, and this is especially true of flexural members, to increase the flange steel area locally. There is a need to consider bond and transfer lengths, as with a reinforcing bar, but there is no special requirement to attach the additional plates to the other flange plates mechanically. By this or related means, variable section properties can be created if this is desired or there is a need. Such additions work well because of the fully enclosed nature of the sections.

6.5 Conclusions

In this chapter a brief introduction has been given to employing some of the earlier mechanics results in the design of actual structural components. Rather than repeat descriptions of construction methods already well established and described elsewhere, a new type of structural material has been described that offers scope to improve on present practices. This material, termed externally reinforced concrete, uses the traditional two materials, steel and concrete, but combines them in a different manner to conventional bar reinforced concrete. The motivation has been to describe a material and method of construction that offers scope to simplify the processes and scope to utilise the concrete material more effectively. For the future we need to consider other options for using our traditional materials if further progress is to be made and safer, more affordable structures are to be available to all our communities, and not just the most affluent and technologically advanced.

There are many facets to the proposals for new construction methods described here. There is scope for readers to make a contribution their own. Important gains can be made if new technology can be proven and later adopted. Whether the technology described here will contribute to meeting future construction needs only time will tell.

6.6 Exercises

(1) The expressions for stiffness and strength given in Section 6.4 are for the simplest case of the simple external case and concrete filling. Write down the additional terms needed to describe the inclusion of a duct in both the stiffness and strength formulae. Analyse the data for a beam-like element suitable for multiple, side-by-side use as a one-way spanning floor when the section is 250 by 200 mm with a 100 by 100 mm duct centred 150 mm from the top. The sheet steel at 2 mm and a yield stress of 350 MPa is to be fabricated in the manner suggested in the worked examples in Section 6.4 to give a 4 mm flange and 2 mm web.

(2) Another option is to pre-stress the members to increase capacity without increase in the overall size of the member. The process is similar to conventional pre-stressing except that the additional option exists to stress the empty case. This can be achieved by stressing the tendon against the case itself. This option is not available in conventional pre-stressing. This can be achieved by tensioning with a tendon placed inside the duct and post-tensioned either before or after the concrete is added. Tensioning the case and tendon before filling with concrete is one method for introducing pre-camber into the beam-like element.

(3) The equations (6.4.0) and (6.4.1) use the symbols T and t. The upper-case T is the *flange* thickness of the (outer) casing. The lower-case t is the *web* thickness. A different use is made of these same symbols on Fig 6.0 where lower-case t is the inner duct steel sheet thickness, and upper-case T the sheet thickness of the (main) outer casing.

In (6.4.1) the symbol x is the depth of the *concrete compression zone*, and y is the equivalent measurement for the tension steel. That is, y is the depth of the *tension zone* from the neutral axis to the (outer) surface of the tension flange.

7 Bibliography and concluding exercises

7.0 Bibliography

There is an extensive amount of material available on the subject of plate theory. Our problem here is one of selection. The first title to be mentioned should perhaps be S. P. Timoshenko's *Theory of Plates and Shells* (McGraw-Hill), first published in 1940, with a second edition (in collaboration with S. Woinowsky-Krieger) in 1959. As the title implies, shells as well as plates are dealt with, but the plates are static elastic plates only. The elastic stability of plates is dealt with in Timoshenko's companion work, *Theory of Elastic Stability* (McGraw-Hill), first published in 1936 and re-edited (in collaboration with J. M. Gere) in 1961. These are both standard works, rather than elementary textbooks.

Another standard work, which has exerted a very strong influence on many subsequent workers in this area of mechanics studies, is A. E. H. Love's *A Treatise on the Mathematical Theory of Elasticity* (Cambridge), first published in two volumes in 1892–3, with subsequent editions in one volume. The last (4th edition, 1927) is still in print and useful to some extent.

More elementary texts, which present plate topics at a level approximately equivalent to that aimed at in this text, include L. G. Jaeger, *Elementary Theory of Elastic Plates* (Pergamon, 1964); C. R. Calladine, *Engineering Plasticity* (Pergamon, 1969, rev. 1985); and R. P. Johnson, *Structural Concrete* (McGraw-Hill, 1967), dealing respectively with elastic, plastic and yield line analysis of plates. All these books provide some exercises for the reader and give answers, but Imperial units may be encountered.

The treatment here has given some priority to the geometry of surfaces. The methods and results discussed in the Appendix are of use beyond plate theory. Other more comprehensive accounts of surface geometry which use the methods of the Appendix are D. E. Rutherford, *Vector Methods* (Oliver and Boyd, 1939 and subsequent editions); K. L. Wardle, *Differential Geometry* (Routledge and Kegan Paul, 1965); C. E. Weatherburn, *Differential Geometry of Three Dimensions*

(Cambridge, 2 vols, 1927, 1930) and T. J. Willmore, *Differential Geometry* (Oxford, 1959).

The Weatherburn volumes, though possibly available only in libraries, are especially useful to the engineer, although all the books were written with the mathematics student in mind. The Rutherford text can also very usefully serve as an aid to the study of the topics in Chapter 1.

The contents of Chapter 2 can be supplemented by reading books with a strength of materials flavour, for example H. Ford, *Advanced Strength of Materials* (Longman, 1963) or S. C. Hunter, *Mechanics of Continuous Media* (Ellis Horwood, 1976). An aspect of Chapter 2 is the choice of coordinates and expressions for the force and displacement quantities in the chosen system. There may be a need to consider the plate equations from the point of view of tensor formulations. The standard reference is A. E. Green and W. Zerna, *Theoretical Elasticity* (Oxford, 1954, 1968). This is a difficult book for engineers, and an alternative is B. Spain, *Tensor Calculus* (Oliver and Boyd, 1953), which takes a more leisurely route through the relevant tensor analysis.

Chapter 3 deals with the classical theory of elastic plates, and the amount of material available for collateral or extended reading is very large. The following titles are suggested, partly to draw the reader's attention to older, now less well-known but thoroughly readable accounts. For example, J. Prescott, *Applied Elasticity* (Longman, 1924, and Dover), is still well worth taking down from the library shelf. So too is E. E. Sechler, *Elasticity in Engineering* (Wiley, 1952). Another out-of-print book, with a good coverage of circular elastic plates is C. E. Turner, *Introduction to Plate and Shell Theory* (Longman, 1965). A comprehensive, more modern book, with some coverage of most relevant topics in elastic plates is R. Szilard, *Plate Theory—Classical and Numerical* (Prentice-Hall, 1974, rev. 2004). Another book is L. H. Donnell, *Beams, Plates and Shells*, (McGraw-Hill, 1976). Anisotropic plates are dealt with in S. G. Leknitskii, *Anisotropic Plates* (Gordon and Breach, 1968).

Possible references to plastic plates (Chapter 4) are also very numerous. Here the distinction must be drawn between plastic plates thought of as isotropic and homogeneous—therefore, in practice, solid metal plates, and nonhomogeneous plates such as reinforced concrete slabs. Each type has associated references. Homogeneous plastic plates are dealt with in P. G. Hodge, *Limit Analysis of Rotationally Symmetric Plates and Shells* (Prentice-Hall, 1963) and the same author's *Plastic Analysis of Structures* (McGraw-Hill, 1959). One of the founders of much of the relevant theory, W. Prager, can be read in his *Introduction to Plasticity* (Addison-Wesley, 1959). Calladine's *Engineering Plasticity* mentioned above can also be usefully read. M. A. Save and C. E. Massonet, *Plastic Analysis and Design of Plates, Shells and Disks* (North-Holland, 1972) is also useful.

The field of non-homogeneous plates and concrete slabs is well provided with references. An interesting work is R. H. Wood, *Plastic and Elastic Design of*

Slabs and Plates (Thames and Hudson, 1961). A later work by L. L. Jones and R. H. Wood, *Yield-line Analysis of Slabs* (Thames and Hudson, 1967) is also worth consulting. A pioneering work is K. W. Johansen, *Yield-line Theory* (Cement and Concrete Association, 1962), an English translation of the 1943 Danish work. A more recent work is R. Park and W. L. Gamble, *Reinforced Concrete Slabs* (Wiley, 1980).

Brief but readable accounts of yield-line and strip theory are contained in B. P. Hughes, *Limit State Theory of Reinforced Concrete Design* (Pitman, 1971, 1976) and F. K. Kong and R. H. Evans, *Reinforced and Prestressed Concrete* (Nelson 1975, 1980). The strip theory philosophy is given in English translation from the Swedish original in A. Hillerborg, *The Strip Method of Design* (Viewpoint, 1975).

The subject of optimum plates (Chapter 5) is not well represented in book form and for further reading there is still a need to refer to the periodical literature. A convenient single reference is G. I. N. Rozvany, *Optimal Design of Flexural Systems* (Pergamon, 1976), and the list of references included there. A tabulation of moment volume information is contained in *Am. Soc. Civil Eng., Proc.* **99** (EM6) (Dec. 1973), p. 1301. Some material in R. H. Wood, *Plastic and Elastic Slabs and Plates* (Thames and Hudson, 1961) is also relevant.

In addition to works in English there are relevant works in other languages. Just one will be noted, A. Sawczuk and T. Jaeger, *Grenztragfahigkeits Theorie der Platten* (Springer, 1963). This is a comprehensive and balanced account of plastic plates from the various viewpoints, and includes a discussion of supporting experimental work.

A more recent book containing useful material related to plates and slabs is M.P. Nielsen "Limit analysis and concrete plasticity" (Prentice–Hall, 1984). A yet more recent paper that surveys the numerical computation field is K. D. Andersen, E. Christiansen and M. L. Overton "Computing limit loads by minimizing a sum of norms" SIAM J. Sci. Comput. **19**(3) (1998), pp. 1046–1062. The book by E. H. Mansfield, "The bending and stretching of plates", Oxford, 1992, is an up-dated version of an earlier book. It is a thorough, concise introduction to elastic plate theory and discusses a range of topics and methods.

The book "Plastic and Elastic Design of Slabs and Plates", by R. H. Wood (1913–1987), published in 1961, has been noted earlier as an important collection of methods and observations relating to our main themes here. John Allen, when working on his thesis, pointed out the paragraph on p. 332 in the final section, "Suggestions for future research", (b) "Dimensional analysis applied to slabs". There Wood made the explicit suggestion to explore the role of the plate area and the perimeter as important parameters. This is the core idea exploited in what is here termed "The Comparison Method". The suggestion has been there for over forty years, and has not attracted attention earlier. There is the added overlain complication that Wood himself did not appreciate the significance of, or have pointed out to him, the earlier contributions of Walter Schumann in two papers in the same volume of "Quarterly of Applied Mathematics" **16** (1958) at p 61 and 309, where the role of the plate area and the possibility of useful isoperimetric inequalities in plastic plate theory were first explored. Schumann did not return to the subject. William Prager (1903–1980), who founded and edited the QAM, refers only to the first of the two papers in the relevant parts of his "An Introduction to Plasticity" published in 1959. The second of the two papers is probably the more important.

The section of Chapter 4 dealing with what we have termed the *Comparison* method is the most novel of the theoretical topics dealt with in this book. The content is based on: P. G. Lowe, "Conjectures relating to rigid-plastic plate bending" Int. J. Mech. Sci. **30**(5) (1988), pp. 365–370; Ibid–II, Int. J. Mech. Sci. **30**(11) (1988), pp. 869–876; J. D. Allen, I. F. Collins, and P. G. Lowe, "Limit analysis of plates and isoperimetric inequalities", Phil. Trans. Roy. Soc. Lond. A **347** (1994), pp. 113–137, and the references contained in these papers. The Allen et al paper is the main source for proofs of the inequalities on which the Comparison method depends. A useful further reference, relating to optimal plates, is G. I. N. Rozvany and B. L. Karihaloo (Eds.) "Structural Optimization", Kluwer Academic Publishers, Dordrecht, 1988.

The content of Chapter 6 is dealt with in greater detail in P. G. Lowe, "Externally reinforced concrete—a new steel/concrete composite", Trans. Inst. of Professional Engineers New Zealand **19**(1/CE) (1992), pp. 42–48; "Externally reinforced concrete composites", in Advances in Concrete Technology (The American Concrete Institute, Special Publication 171–26, 1997), pp. 551–568; and "Composite materials in concrete construction" Proceedings of the International Seminar, Univ. of Dundee, Thomas Telford (2002), pp. 17–29. There are other references included in these papers. There are patent specifications that are relevant—Australian patent no: 629887 and New Zealand patent no: 227555—See also P. G. Lowe, "Engineering and Education" in H. R. Drew and S. Pellegrino (eds) "New Approaches to Structural Mechanics, "Shells and Biological Structures", Kluwer, 2002, pp. 165–174 and "Challenges for building construction", Australian J. of Sturct. Eng., Inst. of Eng. Aust. (2004), **5**, pp. 145–151.

The Twin Towers and their collapse have been referred to earlier. Essential reading for any study of these structures is the Report FEMA 403, "World Trade Centre Building Performance Study: Data Collection, Preliminary Observations, and Recommendations", Federal Emergency Management Agency, Washington, DC, 2002. There are many technical questions remaining to be studied about these events, from a variety of viewpoints. See the National Institute of Standards and Technology's 10000 page three-year study of the WTC collapse, published in 2005. A useful, semi-technical and comprehensive account, written before their collapse, is A. K. Gillespie's "Twin Towers, The life of NY City's World Trade Centre", Rutgers University Press,1999.

7.1 Notes on the development of structural mechanics

Timoshenko, Southwell and Michell

There is an extensive literature dealing with development of structural mechanics as a whole. Some readers may wish to make a broad study: it is suggested that these readers should consult one of the books, such as the Timoshenko "History . . ." discussed below. These Notes are confined to topics and biographical material of particular relevance to those parts of the subject dealt with in this book. Study shows there has been a close linkage in the development of structural mechanics with theories of material behaviour, such as the theory of elasticity. It is useful to continue this linkage in our discussion here.

A very detailed, now old but still valuable source of historical information is the Todhunter and Pearson three volume work, "A History of the Theory of Elasticity and of the Strength of Materials", (Cambridge, 1886 and 1893). A more modern, compact, yet comprehensive book, that deals with the broad history of structural mechanics from the Renaissance until about the 1940's, is S. P. Timoshenko's "History of the Strength of Materials", (McGraw-Hill, 1953). Timoshenko (1878–1972). wrote many books throughout his long life and this is one of the most valuable. Essentially he had two careers—one in his native Russia and another in the USA. Leaving Russia during the aftermath of the First World War he eventually resumed his academic career at the University of Michigan in 1928. His "Theory of Elasticity" was published in 1934, while he was at Ann Arbor. In 1936 he moved to Stanford. His "Plates and Shells" book, published in 1940, has been very influential. He wrote several other texts, many of which were very successful. The "History . . ." was published when Timoshenko was 75: it is now over fifty years old and continues to offer much of value—see "Biographical Memoirs of Fellows of the Royal Society, London" **19** (1973), pp. 679–694.

R.V. Southwell (1888–1970) is referred to in Chapter 3. He was one of a small number of British engineer/applied mechanics academics who had a working relationship with Timoshenko. Southwell published his "Theory of Elasticity" in 1936. The book was intended as preparation for study of Love's more demanding "..Elasticity . . .". Comprehensive and fine work that it is, Southwell's book was overshadowed by Timoshenko's book published two years earlier. Southwell also made important contributions to experimental mechanics. His method of presentation for experimental results, especially compression member data, and usually referred to as a "Southwell Plot", we adapt for another purpose in 3.7. Later Southwell, aided by but independent from Alexander Thom, an Oxford colleague and his successor in the Engineering Science Chair, developed the "Relaxation" method of computation. This was based on finite difference methods and was directed to solving two dimensional engineering problems. He published three volumes of "Relaxation" studies, in 1940, 1946 and 1956. It was an ingenious method of manual computation but was quickly overtaken by computer-based methods and ultimately by Finite Element Methods. The books still retain some interest, however, for the subject as a whole. They are a record of accurate and carefully reported studies on a range of structural and other mechanics problems—see Lowe, "Engineering Heritage 2000", Second Australasian Conference, IPENZ, Wellington (2000), pp. 175–181 and Trans. Multi-Disciplinary Eng., Inst. Eng. Aust. (2002), GE26, pp. 57–64.

The basic ideas discussed in Chapter 5, relating to optimal structural performance, build on pioneer studies by A. G. M. Michell (1870–1959). He made his contribution in the first decade of the twentieth century. Michell was a lone worker who had a thriving practice as a consulting engineer in Melbourne. He is better known for his equally pioneering and highly successful developments in lubrication theory and practice. His later passion was for his ingenious crankless engine, which depended critically on the lubrication developments. This absorbed years of his active life and much of the financial resource he derived from his lubrication patents. But it lead to utter disappointment. He was a true polymath! His elder brother, J. H. Michell (1863–1940), outstanding as a young academic, was an exact contemporary and peer of A. E. H. Love. They were academic research rivals early in their careers. The younger Michell's paper entitled "The limits of economy of material in frame structures" Phil. Mag., **48**(5) (1904), p. 298, was the pioneer structural optimization paper. It passed almost unnoticed until the 1950's. Since then the subject has received a substantial amount of attention, though even today the ideas are valued only in certain quarters—see P. G. Lowe, Mechanics of Structures and Materials, ed. Grzebieta et al, Balkema, 1997, pp. 35–41.

Nineteenth century developments

Both Love's "..Elasticity . . ." and H. M. Westergaard's "Theory of Elasticity and Plasticity" (Harvard/Wiley, 1952) contain preliminary chapters on the history of these subjects and this includes some parts of structural mechanics. These chapters are valuable since both authors were themselves important contributors to the development of the subjects. There are several earlier contributors whose works are not well known. Henry Moseley (1801–1872) trained through the Cambridge Mathematical Tripos but did not secure a high enough position in the order of merit—7th Wrangler (in the order of merit for the examinations from a cohort of around 100 entrants) 1826—to make a career in mathematics. He did become an academic, however, when as a clergyman he was appointed to the Foundation Chair of Natural Philosophy and Astronomy at King's College in London. He wrote several very pertinent texts on the mechanics of structures, the most comprehensive was his "The Mechanical Principles of Engineering and Architecture", (Longmans, 1843 and 1855). In these he discussed, for example, the hinge collapse of structures such as the arch. Moseley was, with I. K. Brunel and others, intimately involved in the Mechanical Industry sections of "The Great Exhibition of the Industry of all Nations" which was housed in Paxton's "Crystal Palace" exhibition building in Hyde Park, London in 1851.

A much earlier contributor to our general subject area was Edmund Mariotte (1620–1684). In European science Mariotte replaces Hooke as the originator of the Elastic Law. He made many experimental observations and these are fully reported in his books and especially in his Oeuvres, published in1717 and again in 1740. One observation made by Mariotte was on the strength of an edge supported, square, ductile plate transversely loaded to collapse by a concentrated load at the centre point. He observed that the load required to collapse the plate was independent of the side length of the plate, other properties such as the thickness (moment strength) and material being unchanged. This is consistent with the theory of plastic plates as discussed in Chapter 4. Timoshenko has a very full account of Mariotte's contributions, and see also P. G. Lowe, 11th National Conference on Engineering Heritage, (The Institution of Engineers, Australia, Canberra, 2001), pp. 117–124 and Aust. J. Multi-Discipl. Eng., Inst. Eng. Aust. (2004), **2**, pp. 73–81.

The earliest, essentially modern book devoted solely to the mathematical theory of elasticity is generally considered to be "Leçons sur la Théorie Mathématique de l'Élasticité des Corps Solides", by Gabriel Lamé (1795–1870), published in Paris in 1852. The next important work, and the first in German, was "Theorie der Elasticität Fester Körper", by Alfred Clebsch (1833–1872), Leipzig, 1862. Consider, too, that since the Industrial Revolution, the mathematical theory of elasticity as a description of material behaviour has been, and remains, a key tool in structural mechanics and engineering calculation generally. Neither work was re-published. In 1883 Clebsch's book was translated into French by Barré de St. Venant (1797–1886), the leading elastician of the day. The translation is about double the length of the German original, as the result of St. Venant's valuable annotations.

The middle years of the nineteenth century saw many developments that have had enduring influence on subsequent progress in our subject. One of the most significant publications of that period was Sir William Thomson (later Lord Kelvin) and P.G. Tait's " Treatise on Natural Philosophy" (Oxford, 1867). This 727 page book was described on the title page as "Vol.1", but no later volumes were ever published. It was re-published somewhat later by Cambridge University Press as two parts and remained in print for the next half century. The content most relevant to our subject here was their treatment of elastic materials from a mathematical standpoint, and in particular the choice they made of notation for stress and strain. It was not a very good choice. But the authority of the authors, many times referred to as T & T', meant that other workers who needed to describe stress and strain adopted their notation.

The first text in English devoted primarily to the mathematical theory of elasticity was William J. Ibbetson's "An Elementary Treatise on the Mathematical

Theory of Perfectly Elastic Solids with a Short Account of Viscous Fluids" , (Macmillan,1887), a book of 515 pages. Macmillan was the leading scientific publisher of the day. Ibbetson was born in South Australia in 1861. His later schooling was taken at Haileybury, a private ("public") school in the environs of London. He left school before completing the usual full course of study since by the age of sixteen he was totally deaf. He entered Clare College, Cambridge in 1879. There he studied for the Mathematical Tripos, the most demanding English university course of study of the day, and graduated 17th Wrangler (in the order of merit for the examinations from a cohort of around 100 entrants) in 1882. His Tutor (William Mollison, himself Second Wrangler in 1876) is on record as believing that Ibbetson's deafness contributed significantly to his disappointing Tripos result. Ibbetson was not coached, that is specially prepared for the examinations, by any of the leading coaches of the day, such as E. J. Routh (1831–1907), who was then at the height of his powers and whose success rate with the students he coached was legendary. After graduating B.A. in 1883 it appears Ibbetson remained in Cambridge, though he did not have any official post. He began his book in late 1885, two years after he graduated, and the manuscript was in the hands of the printer by early 1887. It seems relevant to draw attention to Ibbetson's work. The motivation for writing the book was probably to supply a text suitable for Mathematical Tripos candidates. This sort of mechanics had been introduced into the Tripos syllabus in 1873.

The book was reviewed in the influential science journal "Nature" in December 1887 only months after publication. It was a longer than average review and praised many features of the book, also pointing out some deficiencies. Most "Nature" reviews were signed but this one was not. Ibbetson was 26 when the book appeared. He died in October 1889, aged 28. There are several obituaries but they do not give any clue as to the cause of death. The book contains many novel features as well as some less successful developments. It was never reprinted. There is a remarkably broad range of topics covered including clear formulations of strain and stress, the elastic material, solution methods including Green's energy and Airy's stress function methods, beams, plates, principal stress trajectories, curvilinear coordinate formulations, with excursions into inelastic topics, discussion about real materials and a brief section on applications in biology. M. E. Gurtin in his authoritative treatment of "Linear elasticity" in volume VIa/2, Handbuch der Physik, (Springer,1972) includes references to Ibbetson. "Bibliotheca Mechanica", by V.L. Roberts and I. Trent, (Jonathan Hill, 1991), is an annotated catalogue of Professor Roberts' extensive personal collection of mechanics-related books. Included is a description of Ibbetson's book and his Australian origins.

The potential popularity of Ibbetson's book was probably much affected by the publication five years later of the first volume of Love's "*A Treatise on the Mathematical Theory of Elasticity*". Ibbetson's book was pioneering in a number of respects. For example, he took a special interest in notation for stress and strain and wrote in his Preface that his own preference was for a (double) suffix notation. He seems to be one of the earliest workers in this general field of mechanics to recognize the need to pay closer attention to the definition of shear strain: one of his definitions of strain components is the modern one that is compatible with the tensor nature of the quantities. In his section 123, "Strain components", he defines shear strains, s_i, consistent with this tensor nature, though as an expedient, and for compatibility with Thomson and Tait, he used their notation in the body of his book. Because he used a single suffix this is not a complete resolution of the question, but elsewhere he does use a double suffix. The index notation was later to become an accepted subject standard. Love himself changed his notation in the second edition of his great work. But even when making this change Love did not define his shear strain with the "1/2" included, and hence employed a system incompatible with an index notation. The same is true of Timoshenko and Southwell, a generation later. Love's change of notation is explained by him in the preface to the Second Edition of his "Treatise" in 1906 as resulting from "unfavourable responses to the notation employed in the first edition". In this and later editions Love makes reference to Ibbetson. The distinguished applied mathematician, E. A. Milne, who wrote Love's obituary in the *London Royal Society Series* **3**(9) (1941), p. 478, drew attention to this notation issue as being one of just two respects in which he (Milne) thought that Love had erred significantly.

It was bold of Ibbetson to undertake such an ambitious project so soon after graduating and his deafness must have made the task that much more difficult. He did not make any major original contribution to the subject though at the age he died this is not surprising. It seems his English contemporaries were not impressed by his writings. Pearson, Love (in his first edition), Timoshenko and Southwell make no reference to Ibbetson. European workers of the period did reference him at the relevant places in their papers and books.

There have been several authors who have written important books in this general field very early in their careers. Robert Hooke, of the Elasticity Law, wrote his "Micrographia"(London 1665) when younger than 30. Alfred Clebsch was 29 when his "..Fester Körper" was published in 1862. E. J. Routh published his very influential Rigid Dynamics in 1860 as a 29 year old, while William Ibbetson completed his book by the age of 26. A. E. H. Love was 29 when the first of his two "..Elasticity.." volumes appeared in 1892. Professor Rodney Hill was a similar age when his timely and very influential book, "Plasticity", (Oxford, 1950), was published.

Irish engineering students in the late nineteenth century would probably have used Benjamin Williamson's "Introduction to the Mathematical Theory of Stress and Strain of Elastic Solids" (Longmans, 1894). This was one of a suite of texts written by Williamson that were, according to the preface, tailored for his classes in Dublin. This slim volume of just 135 pages has a good coverage of relevant topics. W. J. M. Rankine (1820–1872) in his comprehensive "A Manual of Applied Mechanics" (1858) and "A Manual of Civil Engineering" (1862), both highly influential textbooks of the day, discusses mathematical elasticity along with many other topics. In all there were four "A Manual of.." and all remained in print long after Rankine's death. The three most important ones, the above two and the "A Manual of the Steam Engine and other Prime Movers" (1859), were written almost at the rate of one/year, then a gap of seven years to the fourth, "A Manual of Machinery and Millwork", in 1869. Rankine was appointed second holder of the Regius Professorship of Civil Engineering and Mechanics at Glasgow in 1855 and died at the early age of 52, on Christmas Eve, 1872; see Oxford Dictionary of National Biography, 2004, essay by B. Marsden.

Two multi-volume series have been important in the development of applied mechanics, including structural mechanics. The first is the twenty three volumes of the German language series titled "Der Encyklopädie der Mathematischen Wissenschaften", (Leipzig, 1898–1935). Four of these are subtitled *Mechanik*. These four, c700 page volumes, contain 32 self-contained essays by prominent academics of the day, including Karman, Lamb, Love and Mises. The other series is the "Handbuch der Physik", (Springer). This also was a multi-volume series, first published in the 1920's, and was one of the stimuli for a resurgence of interest in mechanics, including structural mechanics. The second edition started publishing in 1958 and, unlike the first edition, is very largely in English. A volume of special interest is VIa/2 "Mechanics of Solids II", (Springer,1972) and especially the very detailed essay by P. M. Naghdi, "The theory of shells and plates". This includes an extensive bibliography.

An earlier era—John Wallis and his structural mechanics achievements

There have been many important and interesting developments of parts of our subject that are now all but forgotten. The knowledge is largely buried in whatever documents remain from those earlier times. And this is despite the efforts of Todhunter, Pearson and others to survey the field from the earliest times. The first of the Exercises set out in this chapter is a case in point. There we describe one of a class of structural mechanics problems described and solved by John Wallis (1616–1703). The simplest of these structural arrangements he appears to have acquired from an earlier treatise on architecture, as described below. He expanded the layout types to a whole class of similar problems. He then carefully describes and illustrates the layouts before solving the statics of the internal forces. These in turn are only a part of his mechanics studies, and in turn only part of his total output from an amazingly productive life. In a number of respects Wallis's contributions to structural mechanics are very significant and are largely unreported in the more recent literature. From his dates we can see that he lived to a great age by the norms of the period. Sir Isaac Newton (1642–1723) was about 26 years younger than Wallis. Newton is far better known and his contributions are vastly better researched. Also Newton had the advantage of coming after Wallis and hence he had knowledge of his older colleague's achievements.

Prior to 1687, the date of publication of Newton's much celebrated book, "Philosophiae Naturalis Principia Mathematica", in which he set down the Laws of Motion and firmly established the part of mechanics we now know as dynamics, there had been very few serious attempts to write books on mechanics. The best known earlier mechanics treatise was Galilei's 1638 published book, often referred to by the abbreviated, English translated title as his "Two Sciences". Though justly celebrated, this work does not contain any quantitative examples of the type expected of modern presentations of the subject. A very important, but far less well known book, is John Wallis's "Mechanica, sive de Motu" published in three parts (in Latin) in 1670 and 1671. This work does contain statements of many structural mechanics problems which are then completely analysed in the modern sense. The work was an ambitious attempt to formulate the laws of motion, and Wallis had some limited success. On the other hand, some of the later content, and especially Part 3, has a masterly discussion of statics and includes the structural mechanics developments we illustrate in exercise 1 below.

He gives a totally correct presentation of what we would term statically determinate structural frameworks, and proceeds through all the solution details of a number of problems of floor structures composed of determinate, planar timber assemblies. Most interestingly these are problems involving members in flexure, as in a plate or slab, rather than a discussion of tension or compression forces as in a truss. This is surprising at first sight but on reflection is logical. Truss-type frameworks require fixings (connectors). These were not available and are indeed still under active development even today. With no metal fixings available, trusses as structural elements were not easily achieved. Timber floor support systems in flexure of the type Wallis studied could on the other hand be built entirely by the joiners of the time, without the need for any fixings, since all the bearing points between members were in compression and needed only to be kept in correct location. This was one of the joiners' craft skills.

Prof. Jardine's recent biography "On a Grander Scale" (Harper-Collins, 2002), of Wallis's younger contemporary, Sir Christopher Wren, makes reference to the construction of one of these timber floor structures in the Sheldonian Theatre at Oxford. She discusses Wallis's role in assisting Wren with the "design". Reproduced is an illustration of a proposed arrangement of support timbers for such a floor, derived probably from a similar, but not identical, illustration in the 1611 English translation of the Italian architect, Sebastiano Serlio's, classic 1537 "Architettura." text. The general idea for constructing the floor as a self-supporting structure in that manner probably did derive from Serlio. But Serlio's illustration is really only a diagram. The precise arrangement of timbers is not entirely clear. For the arrangement illustrated by Jardine there are sixteen members and eight unknown forces of interaction, if under a "uniform" load. The system is also determinate and hence is soluble by statics alone. But neither Serlio, nor anyone else earlier, attempted to apply statics to such an arrangement to calculate the forces of interaction between the support timbers in the floor.

Wallis successfully analysed similar systems much more complicated in terms of the number of members, an example being illustrated in our Exercise 1. He would have had no difficulty dealing with the eight independent unknowns of the Serlio example. His involvement in these problems is fully reported in his 1671 text. He clearly illustrates the determinate nature of his proposals for a whole series of arrangements of timber floor support structures in the accompanying copper plates. The computations to obtain the interacting forces required that he set up the equilibrium conditions. This he did and he then solved the simultaneous algebraic equations obtained.

The arrangement of support timbers illustrated in the Jardine "Wren" biography divides the square shape to be spanned into five bays in each direction. One example dealt with by Wallis, his Figure 248, is the eight bay equivalent layout. This requires forty-nine members and a minimum of twenty eight unknown forces of interaction. Wallis completed the calculations for these forces in closed (exact) form and without error. His numerical results could not be improved upon even with modern computers. Computer arithmetic is represented in decimal form: Wallis's results were all expressed as the ratio of two integers. His results are reported fully in the text and are accurately illustrated in the plates of his 1671 volume. It is then merely a matter of inspection of the results to find the maximum bending moments in the layout. Usually this would be the most important "design" value to know.

Wallis's achievements with these structural problems were referred to by several later authors. These references to Wallis's structural solutions continued to be made at least until the start of the twentieth century, but with the decline in use of, and dependence on, timber for such floor support construction, the subject has disappeared from the present day technical literature. I am not aware of any comparable calculations to those of Wallis in any other published work either in Wallis's time or later. In terms of the theoretical basis for the construction of floor structures, Wallis's contributions must rank as possibly the earliest successful applications.

A 1981 article in the *Scientific American*, July issue, p.126 - 138, by Professors H. Dorn and R. Mark, is titled "The Architecture of Christopher Wren" and subtitled in part, "Wren ..did not exploit principles of theoretical mechanics in designing his buildings". Their discussion is devoted primarily to the Sheldonian Theatre in Oxford and St Paul's Cathedral. St Paul's was Wren's greatest achievement. Wren was born in 1632 and died in 1723. His early achievements were in astronomy and science. The Great Fire of London in 1666 changed his career direction to architecture. The Sheldonian was his first major commission as an architect. The design he evolved certainly pushed at the boundaries of known construction of the day. For example, the clear spans of the Sheldonian ceiling/roof structure were greater at 70 feet than the main spans of the floors in the World Trade Centre at 62 feet. This was in the 1660's. Wren was young and inexperienced, and the structural material was wood. The building fulfilled the purpose for which it was constructed, as a secular hall for degree ceremonies etc., for about a century and a half before major renovations were carried out in the nineteenth century. Accounts vary but it seems that nothing of the original timberwork remains today. The ceiling space was a University book store for books printed in the basement. Warehoused books are a heavy live load to impose on any floor. The original weather surface of the roof would have been some sort of quarried stone, and much heavier than a present-day choice.

There are elements of uncertainty and possible confusion about what sort of ceiling/roof structure was constructed in the Sheldonian. Some writers, including A. Tinniswood in his biography of Wren, "His Invention so Fertile", (Cape, 2001), and Dorn and Mark, state that the structural components of the ceiling/roof structure were the trusses. They do not mention the in-plane structural assembly of timbers in the ceiling. Jardine on the other hand does not mention the trusses. It is probable that both systems co-existed, with the truss-like assemblies functioning to transfer the roof loads to the planar structural timber floor (the ceiling) via the truss posts, thus to the nodes on the horizontal structural grid, and thence by planar bending to the walls.

Dorn and Mark do not mention Wallis. They refer to the scientific environment and founding the Royal Society. Wren and Wallis were friends, colleagues and founder members of the Royal Society. Both were academics in Oxford at the time. They held the two Savilian chairs, Wren in astronomy and Wallis in geometry. Wren is far better known to posterity than is Wallis. Sixteen years older than Wren, Wallis was second only to Newton on some measures. His mathematical achievements on their own earn him an honoured place in the history of the subject. But his range of interests and achievements was far wider than this. As a young man Wallis had studied and succeeded in analysing, precisely, the statics of these Serlio-inspired floor structures: this in the 1640's. What then of Wren's knowledge of contemporary structural theory and design processes? Most Wren biographers are explicit that Wren did consult Wallis for his structural mechanics expertise when designing the Sheldonian. Wallis's demonstrated knowledge of the force and moment analysis of these self-supporting (timber) floors was amazingly complete and was not equalled again for possibly two hundred years. Wren had access to this expertise. Precisely what *use* he made of it we cannot know. In the Sheldonian Wren certainly did achieve a remarkable building from the structural engineering viewpoint. We also know that several of these long span floor assemblies were built, primarily in Holland. The Dorn and Mark assertion that "Wren did not exploit (known) principles of theoretical (structural) mechanics…" does therefore seem to be unjustified.

Finale

Our theme has been plane structural elements, and the typical element studied has been the plate or slab loaded by transverse loads. In most practical situations these loads would be generated by gravity when the plate or slab is horizontal. These same elements could equally be used as vertical panels or wall elements.

The first five chapters are devoted to theoretical modelling of these important structural elements. The most novel material presented is in the second half of Chapter 4 where the method called the "Comparison Method" for plastic bending is described.

Theoretical models are only really of use if there is prospect of building structures in which the theoretical methods can be employed. This is the focus in Chapter 6. There the premise is that construction methods and practices, and the resulting structures, are of such central importance in all societies that improvements must be sought, continually. This is perhaps an uncomfortable position to adopt. But the spectre of the Twin Towers floors collapsing with such devastating consequences adds urgency to the case for on-going and critical review. Further examination of the technical aspects of the Twin Tower vulnerability is an essential part of these on-going processes.

Chapter 6 includes a very brief description of what is termed the "preferred method" of construction using steel and concrete in composite. The core feature of the "preferred method" is the proposal to fabricate rectangular tubular steel members and fill these with concrete to produce the composite, described as "externally reinforced concrete". Bar reinforcement is not required though it could be added. Assembling of entire structures using such ERC components can be demonstrated to be highly efficient in both construction and cost terms. Perhaps even more important, the structures created in this manner are stiffer, stronger and more ductile than the conventional alternatives, and can achieve more of the "ideal" features discussed in Chapter 6 than can the conventional steel/concrete composite material.

The architectural profession a hundred and more years ago embraced reinforced concrete as the building material of the future. Many phases of this employment followed. Some architects looked upon reinforced concrete as offering the scope to return to true trabeated construction principles which had in their view been compromised for long periods by masonry and brickwork as the dominant building materials. Others were excited by the "plastic" or "fluid" aspects of the then new reinforced concrete. These and other possibilities became practical possibilities with the new reinforced concrete material. A useful reference is Peter Collins' 1959 published book, "Concrete — The Vision of a New Architecture", re-issued in 2004 by McGill-Queen's University Press, with lengthy additional preliminaries by the editors, together with other of Collins' relevant papers.

Trabeated construction is a term derived from the Latin word "trabs" meaning a timber beam and describes a beam and column construction system.

Architectural form wishing to "express" the underlying structure, a trabeated system for example, was given fresh scope to achieve this with the arrival of reinforced concrete. Somewhat later structural steel evolved to offer similar scope. Other architectural fashions, especially after the Second World War, gave rise to such concepts as "The New Brutalism" which was a description of cast insitu reinforced concrete construction which displayed the rough, even crudely constructed, exposed columns, slabs and other elements which resulted from the timber formwork used. Plywood has replaced the timber in shuttering but has not overcome many of the objections to the "Brutalism" syndrome.

With Externally Reinforced Concrete there is no discarding of formwork and hence no waste from this source. Close tolerances on the construction are far easier to achieve and hence many of the weaknesses and vulnerabilities of reinforced concrete do not arise. Scope to use a trabeated construction system is a natural outcome. Equally, construction using panels, both as floor and wall components, with scope to pre- or post- stress components, and a range of other features, is possible.

7.2 Further exercises

Here are gathered exercises, some of which relate to topics which could not be included in the relevant chapters through lack of space. They may relate to the content of several chapters. All are intended to develop and test the reader's skills in formulation and problem solution.

(1) To begin this section, what is perhaps the first quantitative solution to a plate type problem will be considered. This example was posed and solved by John Wallis in the third part of his *Mechanica*, published in London in 1671. Even today the problem has some appeal and relevance.

As posed the problem relates to the use of a grid work of (wooden) beams, suitably jointed to fit together and span a square (or rectangular) area which is greater than the span of any of the available timbers.

Consider first a single module of the whole construction (Fig. 7.0). Let the end reactions be denoted by A, B, C, D, and the interactions by a, b, c, d. When loaded with W as shown, calculate the values of these eight unknowns. The point to note is that the system is determinate, and hence statics alone will allow a solution to be found.

Show that

$$d\left[4+6\left(\frac{l}{L}\right)+4\left(\frac{l}{L}\right)^2+\left(\frac{l}{L}\right)^3\right]=W.\frac{L}{l}\left[1+\frac{l}{L}\right]^3.$$

When L/l = 2, for example, deduce that

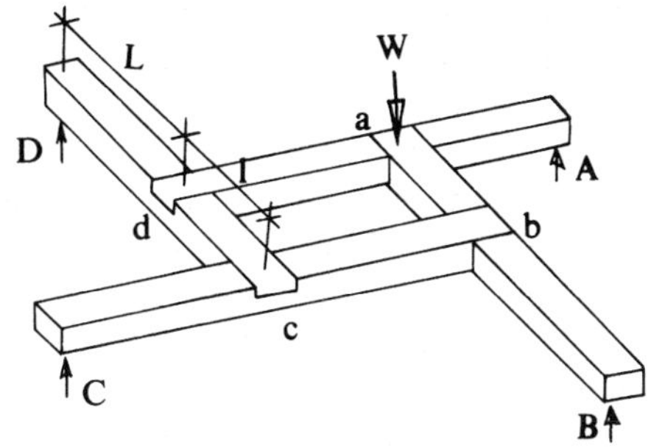

Figure 7.0 The Wallis module.

$$a = 1.246 \text{ W}, \quad b = 0.369 \text{ W}, \quad c = 0.554 \text{ W} \quad \text{and} \quad d=0.831 \text{ W}$$

and finally,

$$D = 0.277 \text{ W}, \quad \text{and} \quad A = 0.415 \text{ W}, \quad B = 0.123\text{W}, \quad C = 0.185 \text{ W}.$$

The Wallis problem is a system composed of an assembly of similar modules (Fig. 7.1). The whole area is a square of side 11L when the main member is of length 3L. There are 40 main members of length 3L, and 20 of length 2L to complete the edge details of the system. Suppose the self-weight of a member of length 3L to be T. Then this is the unit of loading to be assumed in the analysis. The total weight is then $53\frac{1}{3}$ T. The system is assumed to be jointed so that a simple force of interaction is developed at each joint.

By an examination of the layout it can be concluded that the system is statically determinate. With the joint labelling as shown and using symmetry, then the total of 30 unknown forces can be shown to be given by the table of values below.

This problem can be used in various ways. A first procedure might be to check, essentially by back substituting into any relevant static equations, that the results are correct.

Wallis in his 1671 published solution solved for the 25 unknowns (excluding the edge reactions) exactly. If a common denominator of 680 334 is assumed, then all the 25 internal unknowns can be expressed in integral form (Table 7.0). For example, L = 2 513 665/680 334.

(2) Approximate solutions can be of various types. The following is an *elastic* deflection problem for which it is possible to establish upper and lower bounds, by elementary means.

Consider the deflection function

$$w = a(x^2 - 3y^2)^2(x - h)^2.$$

Investigate how the axes are arranged for this expression to describe an elastic equilateral triangular plate with fixed edges. Assume the plate is loaded with a

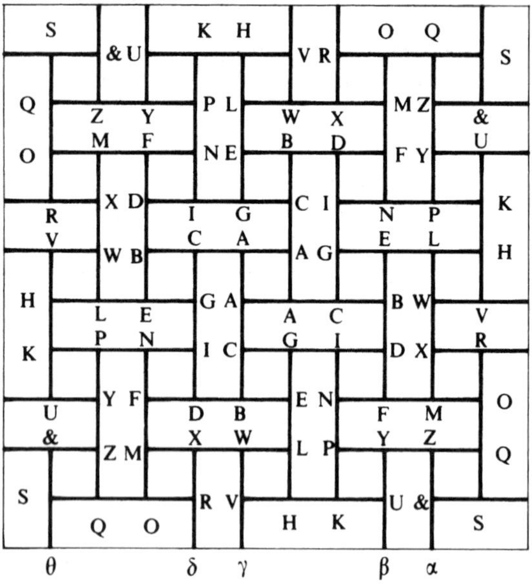

Figure 7.1 Wallis's problem.

Table 7.0 Wallis's problem: solutions

A	9.080	L	3.695	W	3.356
B	8.580	M	3.533	X	2.871
C	7.580	N	7.111	Y	3.212
D	7.185	O	5.572	Z	1.520
E	8.615	P	3.148	&	2.373
F	8.150	Q	2.584	α	1.520
G	7.475	R	5.076	β	3.212
H	5.835	S	4.080	γ	3.357
I	6.290	V	6.046	δ	2.871
K	4.969	U	5.758	θ	2.373

uniform pressure p. Hence show that the maximum displacement w_0 lies between the bounds

$$228 < \frac{w_0 \, . \, 10^6 \, . \, D}{ph^4} < 344.$$

(3) As has been seen in Chapter 4, there are quite a number of exact solutions known for simply-supported plastic plates governed by the square yield locus. There are, in contrast, fewer for clamped-edge plastic plates. A simple one is that for the clamped circular plate. An important, but complicated solution, is that for the square clamped plate, derived by E. N. Fox (*Phil. Trans. Roy. Soc.* (*A*) **277** (1974) 121–155). The main conclusion of this paper is that the collapse pressure, p_c, is given by $p_c . L^2/M_p = 42.851$. By reference to the data in that paper, and suitable interpolations, deduce that A = 3.750, B = 7.074 where A is the area interior to a square of side length 2 which spans the mechanism which develops, and B is the boundary length of this area.

Hence show that

$$\frac{p_c}{M_p}\left(\frac{A}{B}\right)^2 = 3.01$$

for this case.

Though no proof has been offered for the proposition that $p_c A^2/M_p B^2$ should have a lower limit of 3, this result is important additional evidence in support of the inequality $p_c A^2/M_p B^2 \geqslant 3$ for clamped edge plates.

(4) Consider a six-sided simply supported plate with two axes of symmetry such that two opposite sides are parallel and of length b. The remaining four sides are of length a, are at right angles in pairs and form ends of the plate. To check your visualization, when b = 0, the shape becomes a square of side a. Investigate the values of $p_u . A/M_p$, B^2/A and $p_u/M_p . (A/B)^2$ for a mechanism of conical type, as a function of the b/a ratio. Show that all three parameters have minima at the common value of $b/a = 2 - \sqrt{2}$. Show further that the minima are 21.94, 14.60, 1.50, and hence that these are not less than the corresponding values for the circular plate of 18.85, 12.56, 1.50. Note further that when $b/a = 2 - \sqrt{2}$, then the dihedral (maximum) slopes of the slab segments in the mechanism are all equal. Although it will not be proved, this example is of special interest because, though not a fully symmetric shape of plate, $p_u = p_c$, that is the upper bound solution is exact.

(5) Consider a simply supported isotropic plate of arbitrary shape and pressure loaded to collapse. Show that if the mechanism of collapse is pyramid-shaped and has constant slope at the edge, then

$$\frac{p_c}{M_p}\left[\frac{A}{B}\right]^2 = 1.5$$

where A = area, B = perimeter length.

A sketch of a proof is

$$A = \int \frac{\Delta}{\phi}\frac{dB}{2} = \frac{\Delta}{\phi}\frac{B}{2}$$

where Δ = apex deflection, ϕ = edge slope, since ϕ is constant. Hence, the work equation gives $1/3pA\Delta = mB\phi$ and the result follows. Note that a number of results, in Chapter 4 especially, fit this specification.

(6) Space limitations have not permitted inclusion of a discussion of platc dynamics. Hence, as an exercise, confirm that

$$D\nabla^4 w = -m\ddot{w}, \quad (\because) = \frac{\delta(..)}{\delta t}$$

is the relevant dynamic equation, where m is the unit mass of the plate. This equation can be solved, but only with more difficulty than the biharmonic. An approximate procedure which is useful, for example, in exploration of solutions to such equations is the *collocation method.* Essentially, the method aims to satisfy the equation *exactly* at a few chosen points, and not elsewhere, starting from an assumed shape for w, with suitable parameters. Suppose the natural frequency of a square, uniform, simply supported plate of side length a is sought. First assume $w(x,y,t) = A(x,y). \sin \omega_n t$, where ω_n is the frequency sought. Then assume say

$$A(x,y) = A\left[\left(\frac{a}{2}\right)^2 - x^2\right]^2 \left[\left(\frac{a}{2}\right)^2 - y^2\right]^2$$

as a trial shape satisfying the boundary conditions, and just a single unknown parameter A, the amplitude. Hence, show that

$$\frac{m.a^4.\omega_n^2}{D} = 35.7^2$$

is an estimate for ω_n, if the dynamic equation is satisfied at the plate centre.

(7) Show that the information contained in the matrix equation (3.1.3) can be written alternatively as

$$M_{\alpha\beta} = D\left[(1-\nu)\kappa_{\alpha\beta} + \nu.\delta_{\alpha\beta}.\kappa_{\gamma\gamma}\right], \quad (\alpha, \beta = x, y).$$

Here the index notation and repeated indices (1.5) are being used, and δ is the delta symbol, such that $\delta_{\alpha\beta} = 1, \alpha = \beta, \delta_{\alpha\beta} = 0, \alpha \neq \beta$.

(8) Invert (3.1.3) and then show that $\kappa_{\alpha\beta}$ can be written in terms of $M_{\alpha\beta}$ as

$$E.\kappa_{\alpha\beta} = (1+\nu)M_{\alpha\beta} - \nu \,.\, \delta_{\alpha\beta} \,.\, M_{\gamma\gamma}, \quad E = Et^3/12.$$

(9) The potential energy of bending in a plate element is given by

$$\tfrac{1}{2} M_{\alpha\beta} \,.\, \kappa_{\alpha\beta} \,.\, dA \quad (A = \text{plate area}).$$

Hence show that the total potential energy V is given by

$$V = \frac{1}{2}\int M_{\alpha\beta}\kappa_{\alpha\beta} dA - \int pw\, dA$$

where $M_{\alpha\beta}$ are in equilibrium with loads p, $\kappa_{\alpha\beta}$ are derived from w, and $M_{\alpha\beta} = f(\kappa_{\alpha\beta})$ as in question (7) above.

(10) Use the expressions in question (9) to show that

$$V = \frac{D}{2}\int \{[w,_{11} + w,_{22}]^2 - 2(1-\nu)\,[w,_{11}\, w,_{22} - w,_{12}^2\,]\} \,.\, dA - \int pw\, dA.$$

Hence show that if

$$w = A \sin\frac{\pi x}{a} \sin\frac{\pi y}{a}$$

then the maximum displacement (A) for a pressure loaded square simply supported plate is obtained by minimizing V, when

$$A = \frac{4}{\pi^6} \cdot \frac{pa^4}{D}.$$

(11) Chapter 4, Plastic plates, is already the longest chapter in the book, yet we should for completeness include a discussion of the Mises' yield criterion applied to the circular plate. The main features are dealt with here as an exercise. The results are important because a situation arises here which has not been finally resolved and probably denotes one limit on the validity of some aspects of the Comparison Method.

The first essentially exact solution for the axi-symmetric, pressure loaded, clamped edge, circular plate collapsing according to the Mises' criterion was provided by Hopkins and Wang in their paper in *Journal of Mechanics and Physics of Solids* **3** (1954), p. 117. Theirs is an iterative solution, based in part on the equivalent Tresca criterion solution. An independent calculation, not in any way based on the Tresca solution, confirms the accuracy of their solution for the collapse pressure. Thus the parameter γ = p.A/m = 12.53 for the clamped edge case by the independent solution. The simply supported circular plate solution with the Mises' criterion is given by γ = 6.51. These two values compare with the Johansen (Square) yield locus values of 12 and 6, respectively, and 11.25 and 6 for the Tresca material.

The Mises' yield criterion can be written as $M^2 - M.m + m^2 = m_p^2$, where M and m are the principal moments, M radial and m circumferential, and m_p is the full plastic moment/unit length. This is an ellipse in the M,m plane, and circumscribes the Tresca hexagon in similar coordinates. Using the equilibrium equations as described in Chap 2 for the circular plate, set up an integration system and obtain the solution for the collapse pressure. The suggestion is to eliminate the shear force and circumferential moment (m) to give a single, first order differential equation for M which can be integrated over the range r = 0 and $M = m = -m_p$, to r=R when $M=(2/\sqrt{3})^* \; m_p$. The limit at r=0 is a symmetry requirement. At the outer edge, for the clamped condition, M must take on the value to ensure that the normal to the yield locus has no component in the m direction for the zero transverse curvature condition to be achieved. For the simply supported case the conditions are $M = -m_p$ @ r = 0 and M = 0 @ r = R.

Some of the stages on the way might look like, $2(rM)_{,r} = M + \sqrt{(4m_p^2 - 3M^2)} - pr^2$; and when changed to a forward difference scheme for finding M, $(M_{,r})_I = (M_{i+1} - M_i)/h$, with h the interval chosen. An iteration with trial values for the non-dimensional radius $\sqrt{(pr^2/m_p)}$ can quickly converge on the solution sought. On a spreadsheet several hundred iterations can be accommodated and a solution of sufficient accuracy found.

The limitation on the validity of the Comparison method as a source of lower bounds is in question here because when we use the γ= 12.53 value for the Mises circular plate to generate the lower bound for other shapes the method fails for the square, since the Comparison method gives p_c/m_p = 44.46/(side length)^2 where the best available independent calculation, due to Christiansen et al (see reference above) gives a coefficient of 44.13. While the difference is not great in engineering terms it is significant in numerical analysis terms and needs to be investigated. As noted above, this example probably represents one of the boundaries for the validity of the Comparison method.

7.3. Concluding remarks

This is a suitable place, and these exercises are appropriate topics, on which to conclude our discussion of plates and slabs. There are still many avenues left for worthwhile investigation and scope for improvement of the end products: the floor structures in our constructions.

Appendix: Geometry of surfaces

A.0 The need for geometry

The design of a plate as a load-bearing element in a structure or machine requires that the *deformation* of the plate be understood. Initially a plate is plane, or nearly so. When loaded it deflects into some shape which is not plane; hence there is a need to describe this new, curved shape in order to compare with the unloaded plane shape and so evaluate the internal actions generated.

For the purposes of this chapter the plate will be characterized by the "middle surface" in the case of a solid plate, or some other representative surface for composite plates. The middle surface, as the name implies, is the central layer equidistant from the faces of the plate. Thus the deformation of the plate will be characterized by the deformation of the middle surface. This is reasonable, since most plates are thin compared with the distances spanned by the plate and, as already discussed for beams (1.6), the plate will be assumed to obey the Euler—Bernoulli hypothesis of plane sections remaining plane. However, unlike beams whose *geometry* of deformation is comparatively simple, plates are by definition two-dimensional objects and their deformation presents other aspects which can usefully be studied in this chapter.

A.1 Geometry of a plane curve—curvature

A plane curve possesses *shape* by virtue of the curvature of the curve. Consider a small piece of the curve. Assuming this piece to be circular, of radius ρ (see Fig. A.0), then the *curvature* of this portion of curve is defined as $\kappa \equiv 1/\rho$. The quantity ρ is known as the *radius* of curvature.

If the element of arc length is denoted by ds and $d\psi$ is the angle between adjacent radii of curvature then

$$\kappa = 1/\rho = \frac{d\psi}{ds}. \qquad \text{(A.1.0)}$$

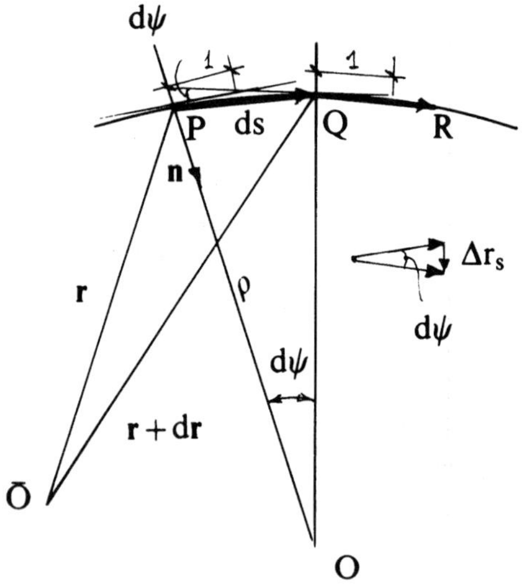

Figure A.0 Geometry of a curve.

Now rework this definition in a vector form. Consider a radius vector **r** to point **P**. Then for Q, the radius vector will be **r** + d**r**. Hence d**r** is a *tangent* vector. Further, d**r**/ds is a *unit* tangent vector (**t**) because |d**r**| = ds. Denote d**r**/ds dy $\mathbf{r}_s$, and consider what meaning can be attached to $\mathbf{r}_{ss}$ (= $d^2\mathbf{r}/ds^2$).

$$\mathbf{r}_{ss} = \frac{(\mathbf{r}_s)_Q - (\mathbf{r}_s)_P}{ds} = \frac{\Delta \mathbf{r}_s}{ds}. \tag{A.1.1}$$

From Fig. A.0 we see that $(\mathbf{r}_s)_Q$ and $(\mathbf{r}_s)_P$ are inclined at dψ and the difference is directed along the inward normal **n**. Comparing (A.1.0) and (A.1.1) it is seen that

$$\kappa = \mathbf{n} \, . \, \mathbf{r}_{ss}, \tag{A.1.2}$$

where **n** is a *unit* inward normal, *directed toward the centre of curvature, O*. This is a convenient vectorial expression for curvature of a plane curve which will also be found useful in the later discussion of surfaces.

Example: Derive an expression for the curvature (κ) of a plane curve in terms of cartesian coordinates (Fig. A.1).

Now

$$\mathbf{r} = x\mathbf{i} + y\mathbf{j},$$

where, **i**, **j** are fixed unit vectors. The curve is a one-parameter object such that y = f(x), say, where f(x) is given.

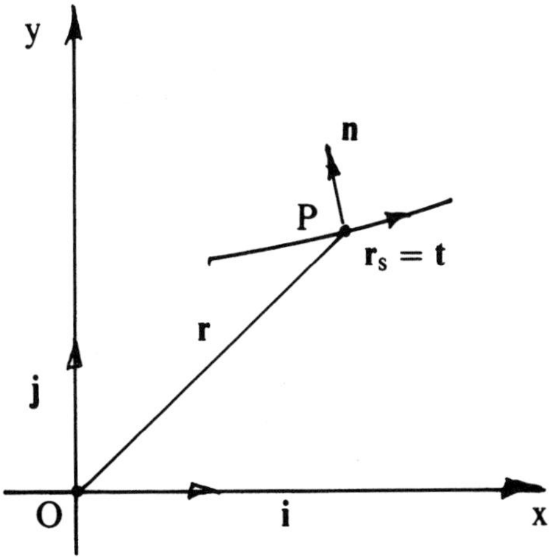

Figure A.1 Plane curve—curvature.

Hence

$$\mathbf{r} = x\mathbf{i} + f(x)\mathbf{j}.$$

Now

$$ds^2 = dx^2 + dy^2 = (1+f_x^2)dx^2,$$

where

$$f_x = \frac{df}{dx}, \quad \text{or} \quad ds = \sqrt{(1+f_x^2)}\,dx.$$

$$\therefore\ \mathbf{r}_s = \frac{d\mathbf{r}}{ds} = \frac{d\mathbf{r}}{dx}.\frac{dx}{ds} = [\mathbf{i} + f_x\mathbf{j}].\frac{1}{\sqrt{1+f_x^2}}, = \text{unit tangent vector.}$$

$$\therefore\ \mathbf{n} = \textbf{unit normal vector} = \frac{-f_x\mathbf{i} + \mathbf{j}}{\sqrt{1+f_x^2}}$$

and note that $\mathbf{n}.\mathbf{t} = \mathbf{n}.\mathbf{r}_s = 0$, since **n** is the normal and $\mathbf{r}_s$ the tangent to the curve.

Then

$$\mathbf{r}_{ss} = \frac{d\mathbf{t}}{ds} = \frac{d\mathbf{t}}{dx}.\frac{dx}{ds} = \frac{(f_x\mathbf{i} + \mathbf{j})f_{xx}}{(1+f_x^2)^2}$$

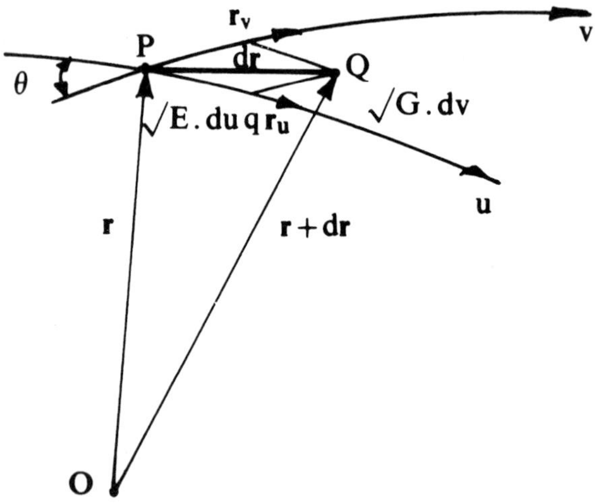

Figure A.2 Surface-element of arc.

Hence

$$\kappa \equiv \mathbf{n}.\mathbf{r}_{ss} = \frac{f_{xx}}{(1+f_x^2)^{3/2}} = \frac{d^2y/dx^2}{\left[1+\left(\dfrac{dy}{dx}\right)^2\right]^{3/2}}.$$

A.2 Length measurement on a surface—first fundamental form

Before any useful analysis of surface geometry can be undertaken, certain preliminaries must be gone through. The first of these is to discuss the measurement of lengths on a surface. Consider the surface to be covered with a coordinate system of two independent variables u, v. There is no need to be specific about u and v at the moment, except to note that each point on the surface can be identified uniquely by a (u, v) pair of coordinates. Consider a typical point P on the surface (Fig. A.2), and a neighbouring point Q. Choose a fixed origin, O, and let the point P be located by the radius vector ($\mathbf{r}$) from O such that

$$\mathbf{r} = \mathbf{r}(u, v) \qquad (A.2.0)$$

Distances measured along the surface, say from P to Q, will be denoted by arc length s, or in this particular case by ds since the points, P, Q are regarded as infinitesimally separated. From the vector relations indicated on Fig. A.2 it is seen that ds is the modulus of d**r**. In order to evaluate ds the standard approach is to form

$$ds^2 = d\mathbf{r} \,.\, d\mathbf{r}. \tag{A.2.1}$$

But

$$d\mathbf{r} = \mathbf{r}_u\, du + \mathbf{r}_v\, dv \tag{A.2.2}$$

since $\mathbf{r} = \mathbf{r}(u, v)$, and where

$$\mathbf{r}_u \equiv \frac{d\mathbf{r}}{du}, \mathbf{r}_v \equiv \frac{d\mathbf{r}}{dv}.$$

Hence

$$\begin{aligned} ds^2 &= (\mathbf{r}_u\, du + \mathbf{r}_v\, dv).\,(\mathbf{r}_u\, du + \mathbf{r}_v\, dv), \\ &\equiv E\, du^2 + 2F\, du\, dv + G\, dv^2, \\ &\equiv I. \end{aligned} \tag{A.2.3}$$

Here the following definitions have been made:

$$E \equiv \mathbf{r}_u \,.\, \mathbf{r}_u, \quad F \equiv \mathbf{r}_u .\, \mathbf{r}_v = \mathbf{r}_v \,.\, \mathbf{r}_u, \quad G \equiv \mathbf{r}_v \,.\, \mathbf{r}_v. \tag{A.2.4}$$

The expression (A.2.3) is known as the *First Fundamental Form* (I), and E, F, G, as the *coefficients* of this form. These definitions and (A.2.3) are very important and can be interpreted as follows. Suppose $dv \equiv 0$. Then Q is at q and the change du produces an arc length Pq which is given by $ds = \sqrt{E}\, du$, from (A.2.3.). These are scalar quantities. But $\mathbf{r}_u$ is a vector, and a *tangent* vector to the surface at P along the direction u increasing, namely P to q. Similarly $\mathbf{r}_v$ is a tangent vector to the surface at P but along the direction v increasingly, as indicated in Fig. A.2.

This is not too surprising, but it is very important: to repeat, r_u, r_v are *tangent* vectors along the u, v lines respectively. Hence d**r**, obtained when both u and v vary to a point such as Q, is itself a tangent vector and relates to r_u, r_v via (A.2.2). These tangent vectors however are often *not* unit vectors. Roughly speaking they are seldom unit vectors when the u, v lines are curved in any sense. The reason is that such lack of straightness usually means that u, v are not a naturally chosen to be arc lengths and this destroys the unit quality.

Consider the cylinder, Fig. A.3. Choose u, v to be arc lengths around and along the cylinder, then

$$\mathbf{r} = 0 \,.\, \mathbf{i} + v\mathbf{j} + R\mathbf{k}$$

and

$$\frac{d\mathbf{i}}{d\theta} = -\mathbf{k}, \frac{d\mathbf{k}}{d\theta} = \mathbf{i}, \frac{d\mathbf{j}}{d\theta} = 0.$$

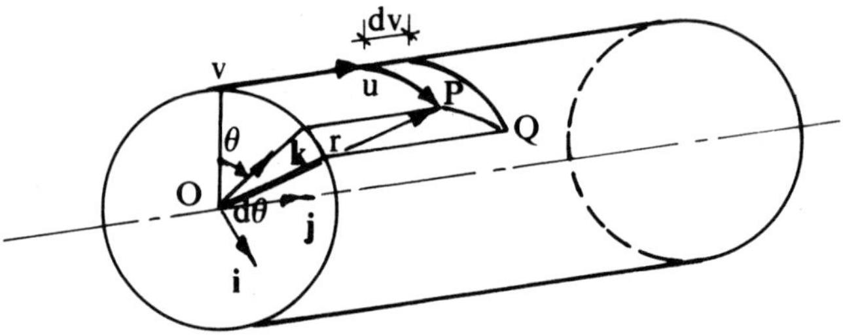

Figure A.3 Geometry on the cylinder (u = Rθ, du = R dθ).

Hence

$$d\mathbf{r} = \overrightarrow{PQ} = R\,d\theta\,.\mathbf{i} + dv.j + 0.\mathbf{k}.$$

But

$$\mathbf{r}_u = \frac{1}{R}.\mathbf{r}_\theta = \frac{1}{R}.[0.-k + 0.\mathbf{j} + R.i],$$
$$= \mathbf{i} \quad \text{and} \quad \mathbf{r}_v = \mathbf{j}$$

Hence

$$E \equiv \mathbf{r}_u \cdot \mathbf{r}_u = \mathbf{i}\cdot\mathbf{i} = 1, \quad F \equiv \mathbf{r}_u \cdot \mathbf{r}_v = \mathbf{i}\cdot\mathbf{j} = 0, \quad G \equiv \mathbf{r}_v \cdot \mathbf{r}_v = \mathbf{j}\cdot\mathbf{j} = 1.$$

In this case, with this choice of coordinates, E = G = 1, F = 0. However, suppose u = θ, with v, as before, the (axial) arc length (Fig. A.3). Then

$$\mathbf{r}_u = \mathbf{r}_\theta = R\mathbf{i}, \quad \mathbf{r}_v = \mathbf{j} \quad \text{(as before)}.$$

Hence

$$E = \mathbf{r}_u \cdot \mathbf{r}_u = R^2, \quad F = 0, \quad G = 1.$$

Hence E ≠ G, though F = 0 still.

Yet other coordinates might be chosen, such as a helix and an axial length. In such cases, E, F and G take on non-zero, and probably more complicated, values.

Exercise: Consider the surface z = Kxy, where x, y, z are cartesian axes (that is all three axes are straight and mutually perpendicular). If x, y are chosen to be the independent coordinates, show that the coefficients of the first fundamental form are $E = 1 + (Ky)^2$, $F = K^2xy$, $G = 1 + (Kx)^2$.

The surface z = Kxy is known as a *hyperbolic paraboloid* (H.P.).

The discussion thus far allows quite general choice of u, v coordinates. In the examples above the coordinates were *orthogonal*, and the F = 0, since $\mathbf{r}_u$ and $\mathbf{r}_v$ are orthogonal. If ϕ is the angle between the u, v lines, then by definition of the various quantities,

$$\mathbf{r}_u . \mathbf{r}_v \equiv F = \sqrt{EG}.\cos\phi.$$

Hence

$$\sin\phi = \sqrt{1-\cos^2\phi}$$
$$= \sqrt{1-\frac{F^2}{EG}} = \frac{H}{\sqrt{EG}}, \quad \text{where} \quad H^2 \equiv EG - F^2.$$

Exercise: For the H.P. show that $H^2 = 1 + K^2.(x^2 + y^2)$

A.3 The normal to a surface

At every point P on a smooth surface there exist at least two tangent vectors, for example $\mathbf{r}_u$ and $\mathbf{r}_v$. The plane of these two vectors is the *tangent plane* at P in which all such tangent vectors lie. The vector *normal* to this plane at P is the normal to the surface at P. It is convenient to define this normal vector to be of unit length as follows:

$$\mathbf{n} = \frac{\mathbf{r}_u \times \mathbf{r}_v}{|\mathbf{r}_u \times \mathbf{r}_v|}. \tag{A.3.0}$$

The denominator in this vector product expression is given by

$$\sqrt{EG}\sin\phi = H,$$

since r_u, r_v are inclined at ϕ and $|r_u| = \sqrt{E}, |r_v| = \sqrt{G}$. Since the surface is smooth, every point on the surface possesses a unique normal, the only ambiguity being the *direction* in which $\mathbf{n}$ is considered positive. Once u and v are defined, then (A.3.0) selects the direction for $\mathbf{n}$ positive.

When adjacent points such as P and Q of Fig. A.2 are considered, each has an associated normal. Generally these normals will be oriented in non-parallel directions, though the vector difference between the two normals is infinitesimal by virtue of smoothness. Let $\mathbf{n}$ be the (unit) normal at P, then $\mathbf{n} + d\mathbf{n}$ can be considered to be the (unit) normal at Q. Since $\mathbf{n}$ and $\mathbf{n} + d\mathbf{n}$ are unit vectors, $d\mathbf{n}$ must

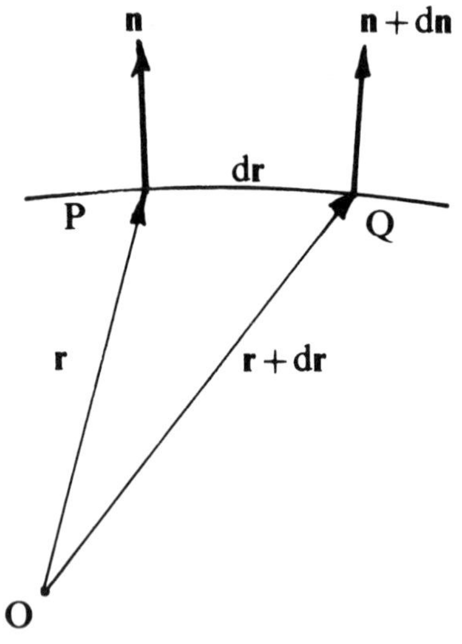

Figure A.4 The surface normal.

be normal to **n** otherwise the length of **n** + d**n** could not be unity. Hence d**n** must lie in the *tangent plane* at P. This is an important conclusion.

Consider the following—what condition must hold if **n** and **n** + d**n** are to be coplanar? Generally these two vectors will be skew. For them to be coplanar then d**n** must lie in the plane of d**r** and **n**.

This latter condition requires that

$$\mathbf{n}.\, d\mathbf{n} \times d\mathbf{r} = 0. \qquad \text{(A.3.1)}$$

Taken a stage further, d**n** and d**r** must be parallel (co-axial) if this condition is to hold, since otherwise d**n** × d**r** will have a component in the **n** direction and the scalar (A.3.1) will not be zero. The next stage is to note that each of d**n** and d**r** is a linear expression in terms of du and dv. Hence (A.3.1) is a quadratic in du and dv. But a ratio of du/dv corresponds to a particular direction of Q in relation to P. Hence (A.3.1) supplies *two specific* directions through P—the *principal directions* at P—for which the normals at neighbouring points *are* coplanar. This topic will be returned to later when the importance of these directions can be further explored.

Exercise: Show that the normal to the hyperbolic paraboloid z = Kxy is

$$\mathbf{n} = (-Ky\mathbf{i} - Kx\mathbf{j} + \mathbf{k}) / \sqrt{1 + K^2x^2 + K^2y^2}\,.$$

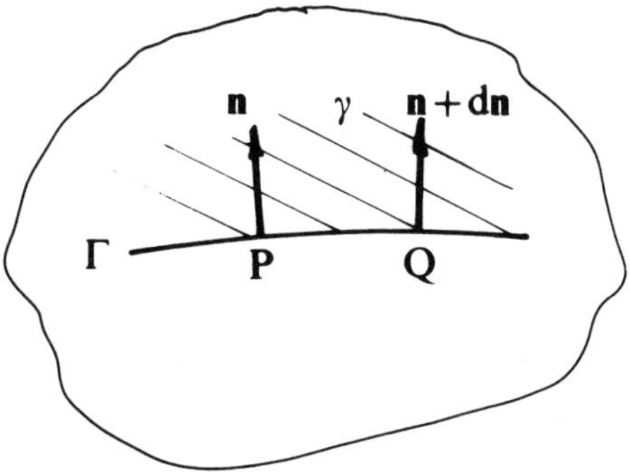

Figure A.5 Surface curvature.

A.4 Normal curvature—second fundamental form

The notion of curvature of a surface is an extension of the notion of curvature of a plane curve. For a plane curve a small piece of curve in the neighbourhood concerned is examined. It is assumed to be accurately described as locally circular and the associated curvature is then the inverse of the radius of this piece of circular curve. Hence *curvature* has the dimensions of L^{-1}.

On a surface, the study is begun by first drawing the normal, **n**, at the point P of interest. But now the curvature is *direction* dependent.

A plane (γ) containing **n** and oriented so as to contain the required direction on the surface is drawn to cut the surface in a piece of (plane) curve Γ (Fig. A.5).

From what was discussed in A.3 there is no guarantee than **n** + d**n**, the unit normal at Q, will lie in γ, unless Γ is a principal direction. Now Γ is regarded as just another piece of plane curve and there will be an associated curvature (κ_p) at P. This is the curvature of the surface at P in the direction P to Q. As Q is chosen to be, in turn, different points in the neighbourhood of P, so κ_p will vary. (How κ_p varies is itself of considerable interest).

From what has already been described, d**n**, the change in **n** away from P, lies parallel to the tangent plane. This is merely a consequence of **n** at all points being restricted to being a *unit* normal. Any change must therefore take place *normal* (transverse) to the existing normal **n**. This point will be pursued in the next section.

Here the aim is to look at some properties of the second derivatives of **r**. Now **n** is normal to each of $\mathbf{r}_u$ and $\mathbf{r}_v$. Hence $\mathbf{n} \cdot \mathbf{r}_u = 0$. Thus it follows that, by taking $(\ )_u$ of this expression,

$$\mathbf{n} \cdot \mathbf{r}_{uu} + \mathbf{n}_u \cdot \mathbf{r}_u = 0,$$

where

$$()_u \equiv \frac{\partial()}{\partial u}, \quad ()_{uu} \equiv \frac{\partial^2()}{\partial u^2}.$$

Rewriting this expression, the normal component of $\mathbf{r}_{uu}$ will be denoted by L when

$$L \equiv \mathbf{n} \,.\, \mathbf{r}_{uu} = -\mathbf{n}_u \,.\, \mathbf{r}_u.$$

In a similar manner

$$N \equiv \mathbf{n} \,.\, \mathbf{r}_{vv} = -\mathbf{n}_v \,.\, \mathbf{r}_v. \tag{A.4.0}$$

Finally

$$M \equiv \mathbf{n} \,.\, \mathbf{r}_{uv} = -\mathbf{n}_u \,.\, \mathbf{r}_v = -\mathbf{n}_v \,.\, \mathbf{r}_u.$$

These three components L, M, N are known as the *coefficients of the second fundamental form* (II).

The second fundamental form itself is defined as

$$II \equiv L\,du^2 + 2M\,du\,dv + N\,dv^2. \tag{A.4.1}$$

The relationship between **r**, **n** and II is then seen to be

$$\begin{aligned} -d\mathbf{n} \,.\, d\mathbf{r} &= -(\mathbf{n}_u\,du + \mathbf{n}_v\,dv) \,.\, (\mathbf{r}_u\,du + \mathbf{r}_v\,dv) \\ &= -(\mathbf{n}_u \,.\, \mathbf{r}_u\,du^2 + (\mathbf{n}_u \,.\, \mathbf{r}_v + \mathbf{n}_v \,.\, \mathbf{r}_u)\,du\,dv + \mathbf{n}_v \,.\, \mathbf{r}_v\,dv^2), \\ &= L\,du^2 + 2M\,du\,dv + N\,dv^2, \\ &= II. \end{aligned} \tag{A. 4.2}$$

Exercise: For the H. P., z = Kxy, show that

$$L = N = O, \quad M = K/\sqrt{1 + K^2(x^2 + y^2)}.$$

By extension from the curvature definition for a plane curve, (A.1.2), the *normal curvature* of the surface can be defined to be

$$\kappa_n \equiv \mathbf{n} \,.\, \mathbf{r}_{ss}. \tag{A.4.3}$$

Now

$$\mathbf{r}_s = \mathbf{r}_u \,.\, u_s + \mathbf{r}_v \,.\, v_s,$$

and

$$\mathbf{r}_{ss} = \mathbf{r}_{uu} \,.\, u_s^2 + 2 \,.\, \mathbf{r}_{uv} \,.\, u_s v_s + \mathbf{r}_{vv} \,.\, v_s^2 + \mathbf{r}_u u_{ss} + \mathbf{r}_v v_{ss}.$$

Hence

$$\kappa_n = \mathbf{n} \,.\, \mathbf{r}_{ss} = Lu_s^2 + 2M \,.\, u_s.v_s + N \,.\, v_s^2,$$
$$= \frac{II}{I}. \qquad \text{(A.4.4)}$$

This is an important basic form for calculation of κ_n.

A.5 The derivatives of n—the Weingarten equations

If d**n** is parallel to the tangent plane, so too are $\mathbf{n}_u$ and $\mathbf{n}_v$.

Hence

$$\mathbf{n}_u = a\mathbf{r}_u + b\mathbf{r}_v \qquad \text{(A.5.0)}$$

where a, b are coefficients to be found.

Now

$$-L = \mathbf{n}_u \,.\, \mathbf{r}_u = a\mathbf{r}_u \,.\, \mathbf{r}_u + b\mathbf{r}_u \,.\, \mathbf{r}_v = Ea + Fb,$$

from the definition of L.

Again

$$-M = \mathbf{n}_u \,.\, \mathbf{r}_v = a\mathbf{r}_v \,.\, \mathbf{r}_u + b\mathbf{r}_v.\, \mathbf{r}_v = Fa + Gb.$$

Hence, here are two equations for a and b. Solve to obtain

$$\begin{bmatrix} a \\ b \end{bmatrix} = \frac{-1}{H^2}\begin{bmatrix} G & -F \\ -F & E \end{bmatrix}\begin{bmatrix} L \\ M \end{bmatrix}.$$

Hence

$$a = \frac{1}{H^2}[FM - GL], \qquad b = \frac{1}{H^2}[FL - EM],$$

where $H^2 = EG - F^2$.

In a similar manner, $\mathbf{n}_v$ can be evaluated.

Finally

$$\mathbf{n}_u = \left(\frac{FM - GL}{H^2}\right)\mathbf{r}_u + \left(\frac{FL - EM}{H^2}\right)\mathbf{r}_v,$$
$$\mathbf{n}_v = \left(\frac{FN - GM}{H^2}\right)\mathbf{r}_u + \left(\frac{FM - EN}{H^2}\right)\mathbf{r}_v. \qquad \text{(A.5.1)}$$

These two equations are known as the Weingarten equations. They will be used extensively in the discussion of curvature to follow.

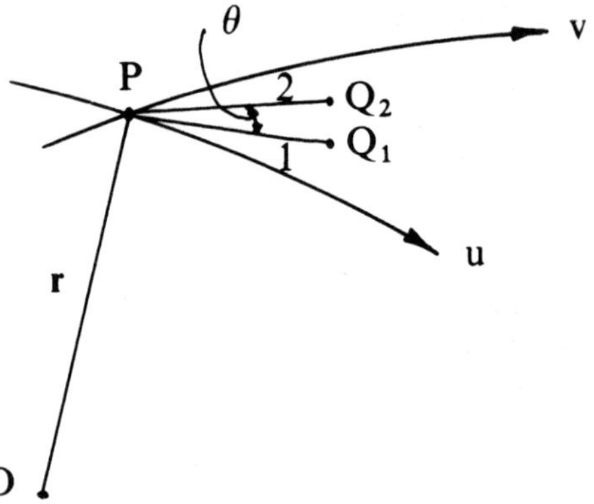

Figure A.6 Directions on the surface.

A.6 Directions on a surface

Consider a typical point, P, on a smooth surface (Fig. A.6), and the u, v coordinate curves through this point. Suppose there are two directions, 1, 2 through P to be investigated. Let PQ_1 be denoted by **dr** and PQ_2 by δ**r**.

Then

$$\mathbf{dr} = \mathbf{r}_u \, . \, du + \mathbf{r}_v \, dv,$$

$$\delta\mathbf{r} = \mathbf{r}_u \delta u + \mathbf{r}_v \delta v.$$

Hence

$$\mathbf{dr} \, . \, \delta\mathbf{r} = E \, du \, . \, \delta u + F(du \, . \, \delta v + dv \, . \, \delta u) + G \, dv \, . \, \delta v.$$

If the angle between these two directions is θ, then ds. δs cos θ = **dr** . δ**r** and the two directions will be *orthogonal* if cos θ = 0. In this case

$$E\left(\frac{du}{dv} \cdot \frac{\delta u}{\delta v}\right) + F\left(\frac{du}{dv} + \frac{\delta u}{\delta v}\right) + G = 0 \qquad \text{(A.6.0)}$$

Now the quantities du/dv, δu/δv essentially describe directions and hence (A.6.0) is the condition that two such directions should be orthogonal.

Exercises: Show that the angle (α) between the tangent vectors $\mathbf{r}_x$, $\mathbf{r}_y$ for the H.P. is in general not π/2. Prove that tan α = H/K^2xy, where

$$H^2 = 1 + K^2(x^2 + y^2).$$

A.7 The principal curvatures

Earlier, in (A.2), it was shown that in general the normals drawn at adjacent points on a smooth surface are non-coplanar. A condition was proposed as a test for the situation when they *are* coplanar. This condition is that **r** and **n** are such that

$$\mathbf{n} \,.\, \mathrm{d}\mathbf{n} \times \mathrm{d}\mathbf{r} = 0. \qquad \text{(A.7.0)}$$

It can be concluded that for (A.7.0) to hold then d**n** and d**r** must be parallel. If not, their vector product will be non-zero and this will point in the **n** direction since d**n** and d**r** lie in the tangent plane. Now, if d**n** and d**r** are parallel, then

$$\mathrm{d}\mathbf{n} = \text{scalar} \,.\, \mathrm{d}\mathbf{r}. \qquad \text{(A.7.1)}$$

Since **n** is dimensionless, being a direction, and **r** has dimensions of length, being a *position* vector, then the scalar must have dimensions of L^{-1}. Indeed, the scalar is in the nature of a curvature. It will now be shown that it is the surface curvature in the direction d**r**, which in this case is a principal curvature. Calling this curvature κ_p, then

$$\kappa_p \,.\, \mathrm{d}\mathbf{r} + \mathrm{d}\mathbf{n} = 0. \qquad \text{(A.7.2)}$$

This equation is known as *Rodrigues' formula*.

Now from (A.4.4)

$$\kappa_n = \frac{\mathrm{II}}{\mathrm{I}} = \frac{-\mathrm{d}\mathbf{n} \,.\, \mathrm{d}\mathbf{r}}{\mathrm{d}s^2}.$$

By taking d**r** of (A.7.2) it is seen that

$$\kappa_p \,\mathrm{d}\mathbf{r} \,.\, \mathrm{d}\mathbf{r} \;(= \kappa_p \,\mathrm{d}s^2)$$

$$= -\mathrm{d}\mathbf{n} \,.\, \mathrm{d}\mathbf{r}.$$

Hence $\kappa_p = \kappa_n$, in this case.

But

$$\mathrm{d}\mathbf{r} = \mathbf{r}_u \,\mathrm{d}u + \mathbf{r}_v \,\mathrm{d}v,$$

and

$$\mathrm{d}\mathbf{n} = \mathbf{n}_u \,\mathrm{d}u + \mathbf{n}_v \mathrm{d}v.$$

Also from the Weingarten equations

$$\mathbf{n}_u . \mathbf{r}_u = \left(\frac{FM - GL}{H^2}\right) E + \left(\frac{FL - EM}{H^2}\right) F = -L.$$

and

$$\mathbf{n}_v.\mathbf{r}_u = \left(\frac{FN - GM}{H^2}\right)E + \left(\frac{EM - EN}{H^2}\right)F,$$

$$= \frac{EFN - EGM + F^2M - EFN}{EG - F^2} = -M.$$

Now take $\mathbf{r}_u$. of (A.7.2) when

$$(\kappa_p E - L)\, du + (\kappa_p . F - M)\, dv = 0.$$

Further, take $\mathbf{r}_v$. of (A.7.2) when

$$(\kappa_p F - M)\, du + (\kappa_p . G - N)\, dv = 0.$$

These two homogeneous equations in (du, dv) imply that

$$\begin{vmatrix} (\kappa_p.E - L) & (\kappa_p.F - M) \\ (\kappa_p.F - M) & (\kappa_p.G - N) \end{vmatrix} = 0,$$

or

$$H^2\kappa_p^2 - \kappa_p(EN - 2FM + GL) + T^2 = 0 \qquad \text{(A.7.3)}$$

where $H^2 = EG - F^2, \quad T^2 = LN - M^2.$

The two roots, κ_p, provided by (A.7.3) are the *two principal curvatures*.
Exercise: For the H. P., z = Kxy, show that the principal curvatures are given by $\kappa_1 = K/H^3, \quad \kappa_2 = -K/H$.

A.8 Principal directions

Along any general direction du/dv the (normal) curvature κ_n is given by

$$\kappa_n = \frac{II}{I} = \frac{L\,du^2 + 2M\,du\,dv + N\,dv^2}{E\,du^2 + 2F\,du\,dv + G\,dv^2},$$

$$= \frac{L\,.\,\lambda^2 + 2M\lambda + N}{E\lambda^2 + 2F\lambda + G}. \qquad \text{(A.8.0)}$$

This curvature has turning values of κ_n when $\lambda(= du/dv)$ takes on the values implied by

$$\frac{d\kappa_n}{d\lambda} = 0 = \frac{(E\lambda^2 + 2F\lambda + G) \,.\, (2L\lambda + 2M) - (L\lambda^2 + 2M\lambda + N) \,.\, (2E\lambda + 2F)}{(E\lambda^2 + 2F\lambda + G)^2}.$$

i.e. when the numerator is zero. Hence

$$(FL - EM)\lambda^2 + (GL - EN)\,\lambda + (GM - FN) = 0. \qquad (A.8.1)$$

The roots λ_1, λ_2 of this equation define two directions through point P for which the curvature has extreme values.

Now from (A.8.1)

$$\lambda_1 + \lambda_2 = -\frac{(GL - EN)}{(FL - EM)},$$

and

$$\lambda_1 \,.\, \lambda_2 = \frac{GM - FN}{FL - EM}, \qquad (A.8.2)$$

If these roots for the directions of extremal κ_n are substituted into (A.6.0) it is found that.

$$E\left(\frac{GM - FN}{FL - EM}\right) - F\left(\frac{GL - EN}{FL - EM}\right) + G = 0.$$

This is an identity. Hence it is concluded that the external κ_n directions are orthogonal

It remains now only to identify these extreme values κ_n with the principal curvatures.

The condition for principal directions is (A.7.0), namely

$$\mathbf{n} \,.\, d\mathbf{n} \times d\mathbf{r} = 0.$$

Let

$$\alpha \equiv \frac{FM - GL}{H^2}, \quad \beta \equiv \frac{FL - EM}{H^2}, \quad \gamma \equiv \frac{FM - EN}{H^2}, \quad \delta \equiv \frac{FN - GM}{H^2},$$

then

$$d\mathbf{n} = (\alpha\mathbf{r}_u + \beta\mathbf{r}_v)\,du + (\delta\mathbf{r}_u + \gamma\mathbf{r}_v)\,dv,$$

and

$$d\mathbf{r} = \mathbf{r}_u\,du + \mathbf{r}_v\,dv.$$

Now $$\mathbf{r}_u \cdot \mathbf{r}_v = H\mathbf{n} \quad \text{from (1.1.5).}$$

Hence

$$d\mathbf{n} \cdot d\mathbf{r} = \alpha H\mathbf{n} \cdot du\, dv - \beta H\mathbf{n}\, du^2 + \delta\, H\mathbf{n}\, dv^2 - \gamma H\mathbf{n} \cdot du \cdot dv,$$

$$\mathbf{n} \cdot d\mathbf{n} \cdot d\mathbf{r} = H[-\beta\, du^2 + (\alpha - \gamma)\, du\, dv - \delta\, dv^2] = 0.$$

Hence

$$(FL - EM)\left(\frac{du}{dv}\right)^2 - (FM - GL - FM + EN)\left(\frac{du}{dv}\right) + (FN - GM) = 0,$$

which is equation (A.8.1). Thus the principal directions are extremal.

The following properties of normal curvature κ_n have now been established. Through every point, P, of a smooth surface the curvature in general varies with the direction through the point. There are two orthogonal directions on which the κ_n is extremal, and these are what have been named the principal directions. Their defining property is that adjacent normals at a pair of neighbouring points on a principal direction are co-planar.

For all other directions in general the adjacent normals are skew lines. As a result the surface along these directions has *twist* as well as (normal) curvature. Only along the two orthogonal principal directions is the twist zero.

Exercise: Show that the extreme values of κ_n are given by

$$(\kappa_n)_e = \frac{M + N\lambda}{F + G\lambda} = \frac{L + M\lambda}{E + F\lambda},$$

where λ is a root of (A.8.1).

Thus far the principal directions have not been thought of as being the coordinate directions. Indeed they will only be the coordinate directions if specially chosen. First, however, if the coordinate directions are principal the $F \equiv 0$, since the directions must be orthogonal. In the next section it will be shown that the further requirement that $M \equiv 0$ should also be true.

A.9 Curvature and twist along the coordinate lines

From the available equations, but especially the Weingarten equations (A.5.1), all the required information relating to curvature and twist can be derived.

Consider a typical point P and the u coordinate direction passing through this point. Along this direction u is changing but v is not. This is the defining property. Then the curvature, κ_{uu}, of the u direction is defined as the negative arc rate of

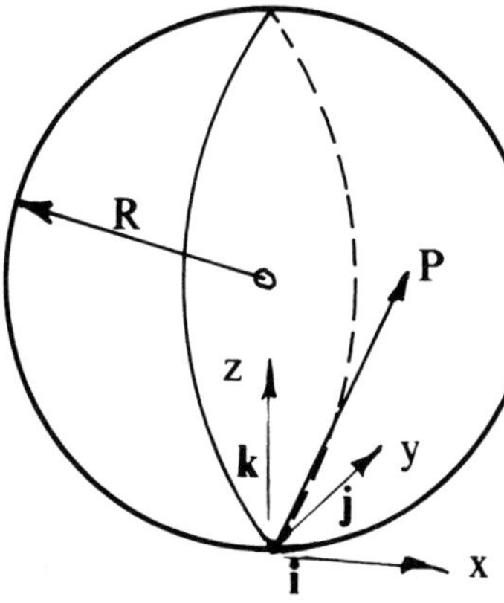

Figure A.7 Geometry on the sphere.

change of **n** in the u direction. This definition can be shown to be equivalent to that proposed earlier (A.1.0, A.1.2, A.4.3).

Expressed in symbols this definition becomes

$$\kappa_{uu} = -\frac{\mathbf{r}_u}{\sqrt{E}} \cdot \mathbf{n}_s$$
$$= -\frac{\mathbf{r}_u . \mathbf{n}_u}{E}, = -\left(\frac{(FM-GL)}{H^2}\right) \cdot \frac{E}{E} - \left(\frac{(FL-EM)}{H^2}\right) \cdot \frac{F}{E}, = \frac{L}{E}. \qquad (A.9.0)$$

If instead of the u direction, the v direction is considered then the curvature, κ_{vv}, of the v line will be found to be

$$\kappa_{vv} = \frac{N}{G}. \qquad (A.9.1)$$

Worked example: Consider the sphere referred to a cartesian coordinated system (Fig. A.7) and evaluate the curvatures κ_{xx}, κ_{yy}.

Now the equation of the sphere referred to these coordinats is

$$x^2 + y^2 + (z - R)^2 = R^2,$$

or

$$z = R - \sqrt{R^2 - (x^2 + y^2)}.$$

Hence at a general point P,

$$\mathbf{r} = x\mathbf{i} + y\mathbf{j} + z\mathbf{k} = x\mathbf{i} + y\mathbf{j} + (R - \alpha)\mathbf{k},$$

where

$$\alpha \equiv \sqrt{R^2-(x^2+y^2)} = f(x, y).$$

Then

$$\mathbf{r}_x = \mathbf{i} + \frac{x}{\alpha}\,.\,\mathbf{k}, \qquad \mathbf{r}_y = \mathbf{j} + \frac{y}{\alpha}\,.\,\mathbf{k}.$$

Hence

$$E \equiv \mathbf{r}_x\,.\,\mathbf{r}_x = 1 + \left(\frac{x}{\alpha}\right)^2, \quad F \equiv \mathbf{r}_x\,.\,\mathbf{r}_y = \frac{xy}{\alpha^2}, \quad G \equiv \mathbf{r}_y\mathbf{r}_y = 1 + \left(\frac{y}{\alpha}\right)^2.$$

Thus

$$\mathbf{n} = \frac{\mathbf{r}_x \times \mathbf{r}_y}{H} = \frac{-(x\mathbf{i} + y\mathbf{j}) + \alpha\mathbf{k}}{R},$$

where

$$H^2 = 1 + \frac{x^2+y^2}{\alpha^2}\,(= EG - F^2) = \left(\frac{R}{\alpha}\right)^2.$$

Then the curvatures κ_{xx}, κ_{yy} can be evaluated as

$$L = -\mathbf{n}_x.\mathbf{r}_x = \frac{\mathbf{i} + \left(\frac{x}{\alpha}\right)\mathbf{k}}{R}.\left(\mathbf{i} + \left(\frac{x}{\alpha}\right)k\right) = \frac{1}{R}\left(1 + \left(\frac{x}{\alpha}\right)^2\right)$$

and

$$\kappa_{xx} = L/E = \frac{1}{R}\frac{\left(1 + \left(\frac{x}{\alpha}\right)^2\right)}{\left(1 + \left(\frac{x}{\alpha}\right)^2\right)} = \kappa(= \text{const.}).$$

In a similar manner

$$N = -\mathbf{n}_y\,.\,\mathbf{r}_y = \frac{\left(\mathbf{j} + \left(\frac{y}{\alpha}\right)\mathbf{k}\right)}{R}.\left(\mathbf{j} + \left(\frac{y}{\alpha}\right)\mathbf{k}\right) = \kappa\left(1 + \left(\frac{y}{\alpha}\right)^2\right)$$

and

$$\kappa_{yy} = N/G = \frac{\kappa\left(1+\left(\frac{y}{\alpha}\right)^2\right)}{\left(1+\left(\frac{y}{\alpha}\right)^2\right)} = \kappa(= \text{const.}).$$

Hence the two curvatures are equal, and each is equal to the (constant) curvature of the sphere—as might have been anticipated.

In addition to the curvatures κ_{uu}, κ_{vv}, the directions u, v in general have associated twists which are also needed for a complete discussion of curvature at a typical point P. The twist at P in the u direction is defined as the negative arc rate of change of **n** *normal* to u. By use of (A.5.1) the appropriate component for twist about u can be seen to be

$$\kappa_{uv} = -\left(\frac{FL - EM}{H^2}\right)\sqrt{G}\,.\,\frac{H}{\sqrt{EG}}\,.\,\frac{1}{E},$$

where now the second term in the $\mathbf{n}_u$ expression has been multiplied by $\sqrt{G}$, the magnitude of $\mathbf{r}_v$, then divided by $\sqrt{E}$ to give the arc rate of change and multiplied by $\sin\theta = H/\sqrt{EG}$ to give the relevant component normal to $\mathbf{r}_u$. After simplification there is obtained

$$\kappa_{uv} = \left(\frac{EM - FL}{EH}\right). \tag{A.9.2}$$

By a similar argument, the twist along v (κ_{vu}) is given by

$$\kappa_{vu} = \left(\frac{GM - FN}{GH}\right). \tag{A.9.3}$$

These twist components can alternatively be shown to represent the arc rates at which n *rotates* about each of u and v, for changes in u and v respectively.

Note that if the coordinate lines (u, v) are orthogonal then F = 0 and

$$\kappa_{uv} = \kappa_{vu} = M/\sqrt{EG}. \tag{A.9.4}$$

Generally, however, for *non-orthogonal* u, v lines, the two twists are not equal. Exceptionally, $\kappa_{uv} = \kappa_{vu}$ also, should L = N = 0. Such surfaces and choice of u, v do occur but are not common.

Exercise: The earlier exercise in this section can now be further examined, for twists. Using the results found there,

$$M \equiv -\mathbf{n}_x \,.\, \mathbf{r}_y = +\kappa\left(\mathbf{i}+\left(\frac{x}{\alpha}\right)\mathbf{k}\right).\left(\mathbf{j}+\left(\frac{y}{\alpha}\right)\mathbf{k}\right) = +\kappa\frac{xy}{\alpha^2}.$$

Hence

$$\kappa_{uv} = \left(\frac{EM-FL}{EH}\right) = \frac{\left[1+\left(\frac{x}{\alpha}\right)^2\right]\times\kappa\frac{xy}{\alpha^2} - \frac{xy}{\alpha^2}\kappa\left(1+\left(\frac{x}{\alpha}\right)^2\right)}{\left[1+\left(\frac{x}{\alpha}\right)^2\right].\kappa/\alpha}$$

$$= 0 = \kappa_{vu}.$$

Hence there are no twists associated with these directions x, y when projected onto the sphere. This result too perhaps could have been anticipated, because of the special nature of the sphere, where every direction is a principal direction.

The defining property for principal directions has been expressed in terms of the normals of adjacent points along the particular direction being *co-planar*. This is equivalent to the requirement that the *twists* should be zero along principal directions.

Now, as noted in A.8, the principal directions are certainly orthogonal and hence $F \equiv 0$. If in addition the associated twists along these directions should be zero, then from (A.9.4) it is seen that this requirement is met if $M \equiv 0$.

Hence the necessary (and sufficient) condition for the *coordinate directions* to be *principal directions* is that

$$F \equiv M \equiv 0 \tag{A.9.5}$$

Exercise: The case of the sphere treated earlier in this section produced $\kappa_{uv} = 0$ for all directions. This is a special case, where *all* directions are principal, even if the two coordinates chosen are not orthogonal.

A. 10 The curvature matrix

In the previous section the curvatures and twists along the u, v coordinate directions were evaluated. In use the four quantities are closely associated; indeed, it is most useful to think of the quantities κ_{uu}, κ_{vv}, κ_{uv} and κ_{vu} as component parts of a single quantity, the curvature matrix κ, which is defined as

$$\kappa \equiv \begin{bmatrix} \kappa_{uu} & \kappa_{uv} \\ \kappa_{vu} & \kappa_{vv} \end{bmatrix}, \tag{A.10.0}$$

$$= \begin{bmatrix} \frac{L}{E} & \left(\frac{EM - FL}{EH}\right) \\ \left(\frac{GM - FN}{GH}\right) & \frac{N}{G} \end{bmatrix}.$$

When the coordinates are general (and $F \neq 0$) this is a non-symmetric matrix. If the coordinates are orthogonal, then $F = 0$ and the matrix becomes symmetric with an off-diagonal term of $M/\sqrt{EG}$. In other words, the curvature matrix for orthogonal coordinates becomes

$$\kappa = \begin{bmatrix} \frac{L}{E} & \frac{M}{\sqrt{EG}} \\ \frac{M}{\sqrt{EG}} & \frac{N}{G} \end{bmatrix}. \tag{A.10.1}$$

The elements in this matrix are the curvatures and twist along two orthogonal directions, u, v, the coordinate directions. If $M \neq 0$ then these directions, though orthogonal, are not principal. Since the coordinate directions u, v are chosen at the outset, usually for convenience, and the principal curvature directions on the deformed plate surface will depend upon how the plate is loaded, it is seldom the case that the u, v directions are principal curvature directions.

Consider an orthogonal choice of coordinates. The question then arises, how can the principal curvature directions through a typical point P be identified, once E, G; L, M, N are known? Now $F = 0$, since the coordinates (u, v) have been chosen to be orthogonal. From (A. 8.1), when $F = 0$, the directions for the principal curvatures are given by the roots of the quadratic

$$EM.\ \lambda^2 + (EN - GL)\lambda - GM = 0. \tag{A.10.2}$$

Hence these directions can be formed, and so too can the principal curvatures from (A.7.3) as the roots of

$$EG\kappa_p^2 - \kappa_p(EN + GL) + T^2 = 0. \tag{A.10.3}$$

If $F \neq 0$, then the general equations (A.7.3) and (A.8.1) must be used.

Exercise: For the H. P., $z = Kxy$, show that when (x, y) are the chosen coordinates then the curvature matrix has the form

$$\kappa = \begin{bmatrix} 0 & \dfrac{K}{H^2} \\ \dfrac{K}{H^2} & 0 \end{bmatrix},$$

i.e. the curvatures are zero, and the twist, in either sense, xy or yx, has the value K/H^2. Thus, although the coordinate lines are non-orthogonal, the curvature matrix is symmetric, but here because of the special circumstance that L = N = 0.

A.11 The curvature circle

The information summarized by the curvature matrix (A.10.0) can for some purposes be more readily represented graphically on what will be called the *curvature circle*.

First recall that at every point, P, on every smooth surface there are just two directions, the principal directions, along which the twist is zero. These directions when extended to adjacent points form a possible system of orthogonal coordinate lines on the surface, called the *lines of curvature* on the surface. Referred to these lines as coordinates, for example lines of latitude and longitude on a perfectly spherical globe, then the curvature matrix becomes a diagonal matrix, since the twist is zero. As noted in A.9, for this choice F = M = 0.

Generally, however, for coordinates which are not lines of curvature, then the curvature matrix will be as shown in (A.10.0). The matrix will be *symmetric* if the coordinates chosen are *orthogonal*, though not necessarily principal.

The curvature circle is drawn on axes of curvature and twist. The circle is centred on the curvature axis (Fig. A.8).

Consider first the case when the coordinate lines (u, v) are orthogonal, though not necessarily lines of curvature. If the coefficients of the fundamental forms are E, G; L, M, N, then

$$\kappa_{uu} = \frac{L}{E}, \quad \kappa_{uv} = \frac{M}{\sqrt{EG}} = \kappa_{vu}, \quad \kappa_{vv} = \frac{N}{G}. \qquad \text{(A.11.0)}$$

Now from (A.10.3) the *principal curvatures* are given by the roots of the quadratic

$$EG\kappa_p^2 - \kappa_p(EN + GL) + (LN - M^2) = 0.$$

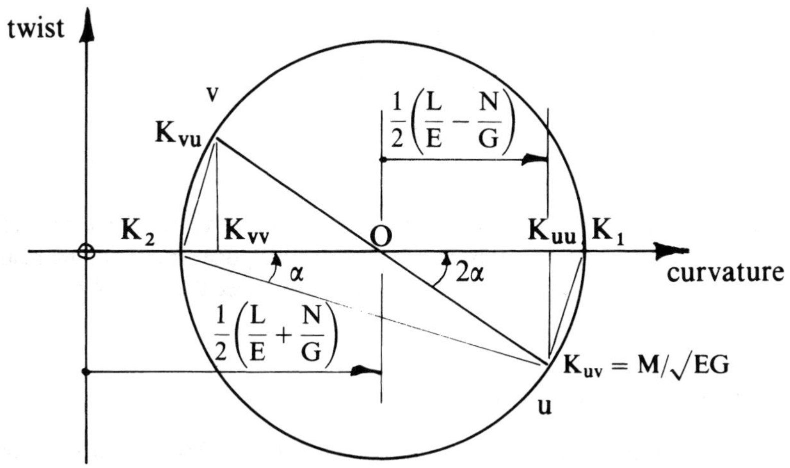

Figure A.8 Curvature circle.

This can be rewritten as

$$\kappa_p^2 - \kappa_p\left(\frac{L}{E}+\frac{N}{G}\right)+\left(\frac{L}{E}\cdot\frac{N}{G}-\left(\frac{M}{\sqrt{EG}}\right)^2\right)=0.$$

Hence, if the principal values are κ_1, κ_2 then

$$\kappa_1+\kappa_2=\frac{L}{E}+\frac{N}{G}, \tag{A.11.1}$$

and

$$\kappa_1\,.\,\kappa_2=\frac{L}{E}\cdot\frac{N}{G}-\left(\frac{M}{\sqrt{EG}}\right)^2.$$

Let

$$4R^2\equiv\left(\frac{L}{E}-\frac{N}{G}\right)^2+4\left(\frac{M}{\sqrt{EG}}\right)^2.$$

Then

$$\kappa_{1,2}=\frac{1}{2}\left(\frac{L}{E}+\frac{N}{G}\right)\pm R. \tag{A.11.2}$$

The relationships (A.11.1) and (A.11.2) are the basis for the circle construction (Fig. A.8). In addition to the principal values it is important to find the inclination of the principal directions to the (u, v) coordinate directions. These inclinations are obtained from (A.10.2) where $\lambda = du/dv$, and the two values are

$$\lambda_{1,2} = \frac{(GL - EN) \pm \sqrt{(GL - EN)^2 - 4.(EM)(-GM)}}{2EM},$$

$$= \frac{\frac{1}{2}\left(\frac{L}{E} - \frac{N}{G}\right) \pm \sqrt{\frac{1}{4}\left(\frac{L}{E} - \frac{N}{G}\right)^2 + \left(\frac{M}{\sqrt{EG}}\right)^2}}{M/G},$$

$$= \frac{\frac{1}{2}\left(\frac{L}{E} - \frac{N}{G}\right) \pm R}{M/G}.$$

But

$$\tan\alpha = \sqrt{\frac{G}{E}} \cdot \frac{dv}{du} = \sqrt{\frac{G}{E}} \cdot \frac{1}{\lambda},$$

from Fig. A.9

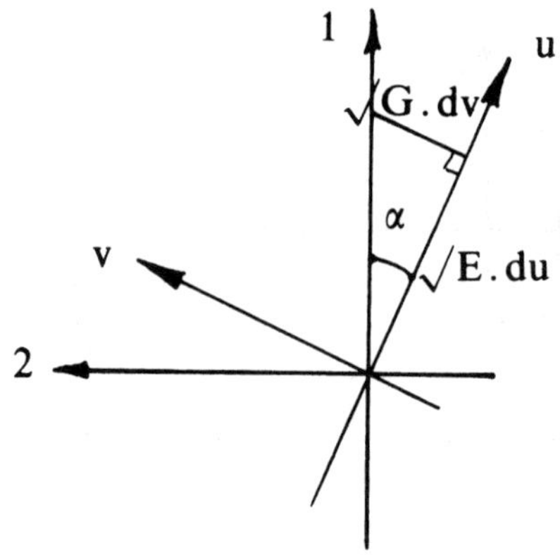

Figure A.9 Principal directions.

$$= \frac{\dfrac{M}{\sqrt{EG}}}{\dfrac{1}{2}\left(\dfrac{L}{E}-\dfrac{N}{G}\right) \pm R}.$$

The plus sign in this expression is associated with the angle shown in Fig. A.8. The minus sign is associated with an angle $[-(\pi/2-\alpha)]$. On the curvature circle it will be noted that with respect to the circle centre, O, an angle of 2α is turned clockwise to arrive at the first principal direction, 1. Alternatively, if an angle $2[(\pi/2)-\alpha]$ from the u direction is turned counter-clockwise then the second principal direction, 2, is obtained.

Thus *directions* in the physical tangent plane of the surface, such as u, v; 1, 2 are represented by *points* on the curvature circle. Further, with centre, O, in the circle, angles *twice* the *physical* angles between particular *directions* must be turned to obtain the corresponding circle *point*. There is direct analogy between the curvature circle and the Mohr circle of stress or strain.

In the general case of non-orthogonal coordinates, then the curvature circle is drawn in slightly different manner. The formulae are now more complicated because there is no symmetry of the curvature matrix, and $F \neq 0$. The circle appears as in Fig. A.10.

The following relations are implied by the circle geometry.

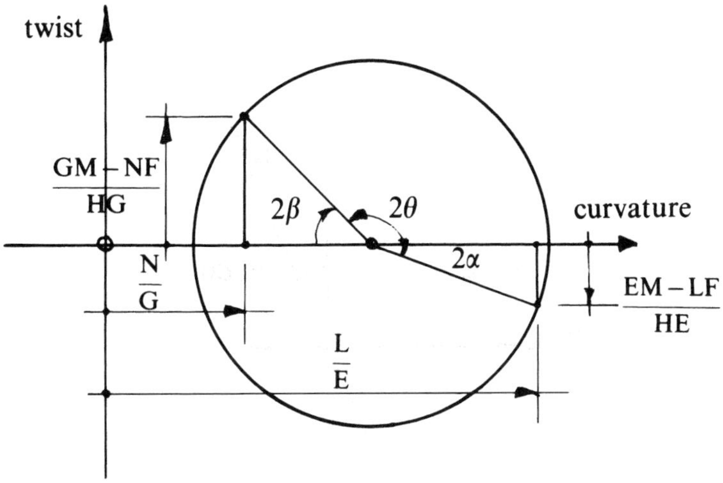

Figure A.10 Curvature circle—non-orthogonal coordinates.

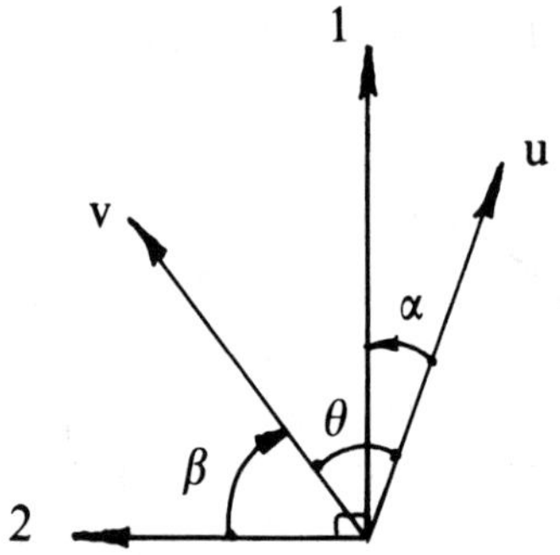

Figure A.11 Principal directions—non-orthogonal coordinates.

$$\beta+\theta=\frac{\pi}{2}+\alpha, \qquad \text{hence} \qquad \cot\alpha=-\tan(\beta+\theta)\ . \tag{A.11.3}$$

If the principal directions are given by the two values of

$$\lambda_i=\frac{d\alpha}{d\beta}(i=1,2), \qquad \text{then}$$

$$\cot\ \alpha=\left(\frac{E\lambda_1+F}{H}\right) \qquad \text{and} \qquad \cot\beta=-\left(\frac{G+F\lambda_2}{H\lambda_2}\right). \tag{A.11.4}$$

Now

$$\tan\theta=\frac{H}{F} \qquad \text{and hence} \qquad \tan(\theta+\beta)=\frac{H}{E\lambda_2+F}. \tag{A.11.5}$$

Finally,

$$E\lambda_1\lambda_2+F(\lambda_1+\lambda_2)+G=0. \tag{A.11.6}$$

This is an identity for the principal directions λ_1, λ_2.
The λ_i values can be obtained from (A.8.1) as the (two) roots of

$$(EM-FL)\lambda^2+(EN-GL)\lambda+(FN-GM)=0. \tag{A.11.7}$$

The principal curvatures are obtainable from (A.8.0) with $\lambda=\lambda_i$ as

$$\kappa_i=\frac{L\lambda_i^2+2M\lambda_i+N}{E\lambda_i^2+2F\lambda_i+G},(i=1,2)\ \cdot \tag{A.11.8}$$

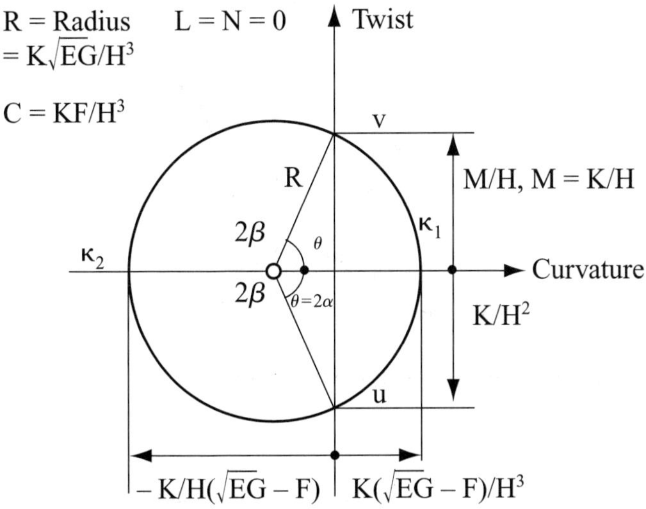

Figure A.12 Curvature circle—hyperbolic paraboloid.

There are various relations between the quantities implied by the circle diagram (Fig. A.10). For example

$$G(EM - FL)\sin 2\beta = E(GM - NF)\sin 2\alpha. \qquad \text{(A.11.9)}$$

Also

$$\sin 2\theta = \frac{2HF}{EG}, \qquad \cos 2\theta = \frac{2F - EG}{EG}, \qquad \text{(A.11.10)}$$

from the definitions of the fundamental quantities.

Another useful relation is

$$E(LG - EN) + 2F(EM - FL) = 2H(EM - FL)\cot 2\alpha. \qquad \text{(A11.11)}$$

The basic results for the coordinate directions u, v are of course the relations

$$\kappa_{uu} = \frac{L}{E}, \quad \kappa_{uv} = \frac{(EM - FL)}{EH}, \quad \kappa_{vu} = \frac{(GM - FN)}{GH}, \quad \kappa_{vv} = \frac{N}{G}. \qquad \text{(A.11.12)}$$

These are the component values occurring in the curvature matrix.

Exercise: For the H. P. $z = Kxy$, with (x, y) chosen as coordinates, show that the curvature circle is given by Fig. A.12.

A.12 Continuity requirements

The core of the study of surface geometry as presented so far depends upon the calculation of six quantities in general, namely E, F, G; L, M, N, the coefficients of the two fundamental forms.

It might be thought that these six quantities are independent; however, they are not. There exist *three* independent relations between the *six* quantities. These relations (the *continuity* requirements) are differential in form and are the requirements that the six quantities should describe a *smooth* surface.

Collectively the continuity requirements are known as the *Codazzi–Gauss equations*; there are two Codazzi equations and one Gauss equation, three equations in all. Should there be some special features about the particular surface and choice of coordinates, then some of the fundamental quantities may be zero and some of the continuity conditions may be identities.

Initially, suppose that the coordinates u, v are *line of curvature* coordinates. The $F = M \equiv 0$. For this choice of coordinates, it is useful to define $A^2 \equiv E$, $B^2 \equiv G$. The A and B are known as the *Lamé parameters*. There are now just four fundamental quantities A, B; L, N.

The three vectors $\mathbf{r}_u$, $\mathbf{r}_v$, $\mathbf{n}$ in this case form a mutually perpendicular system, with $\mathbf{n}$ a unit vector, the normal vector. For definiteness we shall assume $\mathbf{r}_n$, $\mathbf{r}_v$, $\mathbf{n}$ to be a right-handed system. Now evaluate the second derivatives of $\mathbf{r}$ in terms of these three (base) vectors.

From the definitions of L and N, and remembering that M is zero for line of curvature coordinates, then

$$\begin{aligned} \mathbf{r}_{uu} &= L\mathbf{n} + \alpha \,.\, \mathbf{r}_u + a \,.\, \mathbf{r}_v \\ \mathbf{r}_{uv} &= 0 \,.\, \mathbf{n} + \beta \,.\, \mathbf{r}_u + b \,.\, \mathbf{r}_v \\ \mathbf{r}_{vv} &= N\mathbf{n} + \gamma \mathbf{r}_u + c \,.\, \mathbf{r}_v \end{aligned} \qquad \text{(A.12.0)}$$

where α, β, γ; a, b, c are coefficients, to be evaluated.

Now

$$\mathbf{r}_u \,.\, \mathbf{r}_{uu} = \tfrac{1}{2}(\mathbf{r}_u^2)_u = \tfrac{1}{2}E_u.$$

Also

$$\mathbf{r}_v \cdot \mathbf{r}_{uu} = (\mathbf{r}_u . \mathbf{r}_v)_u - \tfrac{1}{2}(\mathbf{r}_u^2)_v = O - \tfrac{1}{2}E_v,$$

together with the similar expressions, $\mathbf{r}_v \cdot \mathbf{r}_{uv} = \tfrac{1}{2}E_v$,

$$\mathbf{r}_v \,.\, \mathbf{r}_{uv} = \tfrac{1}{2}G_u; \quad \mathbf{r}_u \,.\, \mathbf{r}_{vv} = -\tfrac{1}{2}G_u \quad \text{and} \quad \mathbf{r}_v \,.\, \mathbf{r}_{vv} = \tfrac{1}{2}G_v.$$

Then by taking $\mathbf{r}_u$., $\mathbf{r}_v$. in turn of the expressions in (A.12.0) the unknown coefficients α, β, γ; a, b, c can be found.

Thus

$$\mathbf{r}_u \cdot \mathbf{r}_{uu} = \frac{1}{2} E_u = \alpha \cdot E, \quad \text{hence} \quad \alpha = \frac{1}{2E} \cdot E_u,$$

$$\mathbf{r}_v \cdot \mathbf{r}_{uu} = -\frac{1}{2} E_v = aG, \quad \text{and} \quad a = -\frac{1}{2G} E_v. \tag{A.12.1}$$

Similarly,

$$\beta = \frac{1}{2E} E_v, \; b = \frac{1}{2G} G_u; \; \gamma = \frac{1}{2E} G_u, \quad c = \frac{1}{2G} G_v.$$

Consider now the identity

$$(\mathbf{r}_{uu})_v = (\mathbf{r}_{uv})_u.$$

From the knowledge of $\mathbf{r}_{uu}$, $\mathbf{r}_{uv}$ in terms of $\mathbf{r}_u$, $\mathbf{r}_v$, $\mathbf{n}$, this identity can be written in expanded form as

$$L_v\mathbf{n} + L\mathbf{n}_v + \alpha_v\mathbf{r}_u + \alpha\mathbf{r}_{uv} + a_v\mathbf{r}_v + a\mathbf{r}_{uv} = \beta_u\mathbf{r}_u + \beta\mathbf{r}_{uu} + b_u\mathbf{r}_v + b\mathbf{r}_{uv}. \tag{A.12.2}$$

If the expressions for the second derivatives of **r** from (A.12.0) are substituted, and the coefficient of **n** in the resulting expression with a zero right-hand side computed, then this must be zero. So too must the coefficients of $\mathbf{r}_u$ and $\mathbf{r}_v$ be separately zero.

From (A.12.2)

$$(L_v - L\beta + aN)\mathbf{n} + (\;)\mathbf{r}_u + (\;)\mathbf{r}_v = 0.$$

Hence

$$L_v - \beta \cdot L + aN = 0.$$

But

$$L = E\kappa_1, \quad N = G\kappa_2, \quad E = A^2, \quad G = B^2,$$

whence

$$(A^2\kappa_1)_v - \frac{A^2\kappa_1}{2A^2}(A^2)_v = \frac{1}{2B^2}(A^2)_v B^2\kappa_2.$$

Simplifying,

$$(A\kappa_1)_v = \kappa_2 (A)_v. \qquad \text{(A.12.3a)}$$

In a similar manner, by equating the **n** component of $(\mathbf{r}_{vv})_u = (\mathbf{r}_{uv})_v$ to zero there is obtained the equation,

$$(B\kappa_2)_u = \kappa_1 (B)_u. \qquad \text{(A.12.3b)}$$

These two continuity requirements are known as the Codazzi equations. The relations which follow from equating to zero any of the other four components of the vector identities can be shown to give rise to just a single further equation of continuity, the Gauss equation. (The $\mathbf{r}_u$ component of (A.12.2) set to zero leads to an identity, but the $\mathbf{r}_v$ component equated to zero leads to the Gauss equation).

Proceed then to equate to zero the $\mathbf{r}_v$ component of (A.12.2), when there is obtained

$$-\kappa_2 L + \left(\frac{1}{2E} E_u\right) \cdot \left(\frac{1}{2G} G_u\right) + \left(-\frac{1}{2G} E_v\right)_v + \left(-\frac{1}{2G} E_v\right) \cdot \left(\frac{1}{2G} G_v\right)$$
$$= \left(\frac{1}{2E} E_v\right) \cdot \left(-\frac{1}{2G} E_v\right) + \left(\frac{1}{2G} G_u\right)_u + \left(\frac{1}{2G} G_u\right)^2.$$

This equation can be simplied. Now $L = E\kappa_1$, $E = A^2$, $G = B^2$, then there results

$$\left(\frac{1}{A} B_u\right)_u + \left(\frac{1}{B} A_v\right)_v = -AB\kappa_1\kappa_2. \qquad \text{(A.12.4)}$$

This equation of continuity is known as the Gauss equation.

The importance of the Gauss (continuity) equation is that it shows the Gaussian curvature $\kappa_g (\equiv \kappa_1\kappa_2)$ to be a function of A and B only. Note that L, N do not appear in (A.12.4). This is a somewhat surprising, but very important result. Thus it shows that any surface which is *deformed* in such a way that the surface is not stretched, sheared or compressed undergoes *no change* of κ_g at any point, since under these conditions of *inextensional bending* of the surface, neither A nor B for the surface change at any point, and hence the expression for κ_g implied by (A.12.4) remains unchanged.

In addition to the Codazzi—Gauss continuity equations, in some chapters other continuity considerations will arise from the need to ensure slope continuity between adjacent portions of surface. Particularly in the study of optimized plates, there is need to discuss (deflection) surfaces which are made up from portions of different types of surface. A requirement will be that such surfaces should be fully continuous within each portion, that is show *gradual* variation of all quantities such as coordinates, slope and curvature. A composite surface made up from a number of such portions must in addition be coordinate and *slope* continuous across and along the common junction. This requires that pairs of coincident points on the junction, regarded as belonging each to an adjacent portion, should possess common coordinates and common tangent plane. When dealing with plates, which when deflected by load are *shallow* surfaces (see p. 242), such conditions will usually be expressed in terms of displacements and surface slopes.

The details of such problems are dealt with in Chapter 5.

A.13 Special surfaces

Thus far the discussion has considered general surfaces. It is appropriate now to turn to various special classes of surface.

Consider first the class of *surfaces of revolution*. Such surfaces are traced out when a plane curve, coplanar and fixed with respect to an axis, is rotated about this axis. The axis will be referred to as the axis of revolution. If this is taken as the axis for a coordinate, z and d denotes the perpendicular distance to the curve, then the coordinates of a point can be written as

$$x = d(z)\cos\phi, \qquad y = d(z)\sin\phi, \qquad z. \tag{A.13.0}$$

Here x, y are axes fixed in space with an origin on the axis of revolution (z), and ϕ is a longitude angle seen true when the surface is viewed along the z axis. The surface coordinates, u, v will be chosen so that u(= z) varies in the axial direction, and v (= ϕ) varies in the circumferential direction.

Then the position vector of a typical point is given by

$$\mathbf{r} = d\cos\phi\,.\,\mathbf{i} + d\sin\phi\,.\,\mathbf{j} + z\mathbf{k} \tag{A.13.1}$$

where, **i**, **j**, **k** are fixed unit vectors along x, y, z respectively. The surface tangent vectors $\mathbf{r}_z$, $\mathbf{r}_\phi$ can then be found as

$$\begin{aligned} \mathbf{r}_z &= dz\cos\phi\,.\,\mathbf{i} + dz\sin\phi\,.\,\mathbf{j} + k \\ \mathbf{r}_\phi &= -d\sin\phi\,.\,\mathbf{i} + d\cos\phi\,.\,\mathbf{j} + 0 \end{aligned} \tag{A.13.2}$$

Here

$$(\,)_z = \frac{\partial(\,)}{\partial z}, \qquad (\,)_\phi = \frac{\partial(\,)}{\partial \phi}.$$

The coefficients of the first fundamental form are

$$\begin{aligned} E &= \mathbf{r}_z \cdot \mathbf{r}_z = dz^2 \cos^2\phi + dz^2 \sin^2\phi + 1, \\ &= 1 + dz^2, \\ F &= \mathbf{r}_z \cdot \mathbf{r}_\phi = -dz\cos\phi.\sin\phi + dz\sin\phi.\cos\phi = 0, \\ G &= \mathbf{r}_\phi \cdot \mathbf{r}_\phi = d^2. \end{aligned} \tag{A.13.3}$$

From these results it is seen that the coordinate lines are orthogonal, since $F = 0$. Hence $\mathbf{r}_z$ and $\mathbf{r}_\phi$ are orthogonal.

Next the unit normal can be evaluated as

$$\mathbf{n} = \frac{\mathbf{r}_z \times \mathbf{r}_\phi}{(\mathbf{r}_z \times \mathbf{r}_\phi)} = \frac{-d\cos\phi.\mathbf{i} - d\sin\phi.\mathbf{j} + d\,.\,d_z\mathbf{k}}{d\sqrt{(1+dz^2)}}$$

The second derivatives of $\mathbf{r}$ are

$$\begin{aligned} \mathbf{r}_{zz} &= d_{zz}\cos\phi.\mathbf{i} + d_{zz}\sin\phi.\mathbf{j} + 0, \\ \mathbf{r}_{z\phi} &= dz\sin\phi.\mathbf{i} + dz\cos\phi.\mathbf{j} + 0, \\ \mathbf{r}_{\phi\phi} &= -d\cos\phi.\mathbf{i} - d.\sin\phi..\mathbf{j} + 0. \end{aligned}$$

Here the coefficients of the second fundamental form are

$$\begin{aligned} L &\equiv \mathbf{n}.\mathbf{r}_{zz} = \frac{-d_{zz}}{\sqrt{(1+dz^2)}}, \\ M &\equiv \mathbf{n}.\mathbf{r}_{z\phi} = 0, \\ N &\equiv \mathbf{n}.\mathbf{r}_{\phi\phi} = \frac{d}{\sqrt{(1+dz^2)}}. \end{aligned} \tag{A.13.4}$$

Since $M \equiv 0$, and $F = 0$, it follows that the coordinate curves on the surface associated with z, ϕ are lines of curvature.

Hence the principal curvatures are given by

$$\kappa_1 = L/E = -\frac{d_{zz}}{(1+dz^2)^{3/2}} \tag{A.13.5}$$

and

$$\kappa_2 = N/G = \frac{1}{d(1+dz^2)^{1/2}}$$

These expressions can be interpreted as follows.

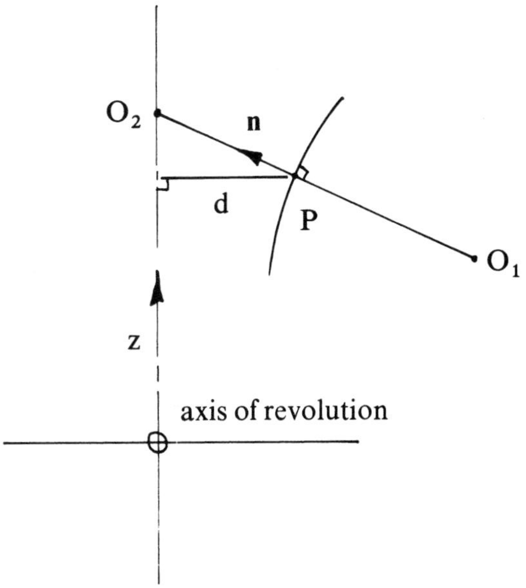

Figure A.13 Surface of revolution—section.

From Fig. (A.13) it can be seen that $\kappa_1 = 1/O_1P$, where O_1 is the centre of curvature for the original plane curve, at the point P. The minus sign relates to **n** being directed *away* from O_1. The point O_2 where the normal intersects to the axis of revolution is the centre of curvature associated with the ϕ coordinate direction since from the geometry of the figure, it can be seen that

$$\kappa_2 = \frac{1}{O_2P}.$$

Here there is a *positive* sign since **n** and O_2 are on the *same* side of the surface.

As a point of interest, for Codazzi–Gauss continuity conditions it can be noted that

$$A = \sqrt{1 + d_z^2}, \quad B = d; \quad u = z, \quad v = \phi, \quad d = d(z).$$

Hence the Gauss equation becomes

$$\left(\frac{1}{A}B_z\right)_z + \left(\frac{1}{B}A_\phi\right)_\phi = -AB\kappa_1\kappa_2.$$

Now

$$\text{L.H.S} = \left(\frac{1}{\sqrt{1+d_z^2}} .dz \right)_z + 0$$

$$= \frac{\sqrt{1+dz^2}\,.\,d_{zz} - dz\dfrac{1.2dz\,.\,d_{zz}}{2\sqrt{1+d_z^2}}}{1+dz^2},$$

$$= \frac{d_{zz}[1+dz^2 - dz^2]}{(1+dz^2)^{3/2}} = \frac{d_{zz}}{(1+d_z^2)^{3/2}}.$$

$$\text{R.H.S} = \frac{-\sqrt{1+dz^2}\,.\,d - d_{zz}}{(1+dz^2)^{3/2}} \cdot \frac{1}{d(1+dz^2)^{1/2}} = \text{L.H.S.}$$

One Codazzi equation only exists because of the symmetry and this becomes

$$(B\kappa_2)_u = \left(d.\frac{1}{d(1+dz^2)^{1/2}} \right)_z = -\frac{dz\,.\,d_{zz}}{(1+dz^2)^{3/2}} = \kappa_1 (B)_u$$

an identity

Exercise: The hyperboloid of revolution is defined by

$$\frac{x^2+y^2}{a^2} - \frac{z^2}{b^2} = 1.$$

If

$$\Omega^2 \equiv b^2 + z^2 \left(1 + \left(\frac{a}{b} \right)^2 \right),$$

show that

$$\kappa_{11} = -b/\Omega a, \qquad \kappa_{22} = ab/\Omega^3,$$

are the principal curvatures.

Another very important class of surfaces of interest is the class of *shallow surfaces*. In fact all the shapes developed by loaded plates, of the type concerned with in this book, are shallow surfaces. By this is meant that the surface *slopes* are everywhere *small*. As an example, this means that a shallow paraboloid of revolution cannot be distinguished from a shallow segmental cap cut from sphere: in each case the shape locally is described by a quadratic variation.

Some surfaces can be thought of as traced out by the translation of one curve over another. Such surfaces are known collectively as *surfaces of translation*.

If the two curves are straight lines, then an example of such a surface is the hyperbolic paraboloid, $z = Kxy$, which has been studied in the earlier exercises.

Certain surfaces have the property that by suitable *unfolding* they can be laid flat on the plane. These are the *developable surfaces*. Examples (amongst others) are cylinders and cones. But the sphere cannot be so unfolded, hence the need for some (agreed) projection (i.e. distortion) of the surface of the sphere earth in order to produce flat maps of a whole or part of the surface. The cylinders and cones when so unfolded are undergoing *inextensional bending*, the type of deformation which neither stretches, shears nor compresses any portion of the surface. As a result, the starting value of the Gaussian curvature, κ_g (= product of the principal curvatures), is maintained. But since the ultimate unfolding on to the plane produces a surface (the plane) which clearly has zero κ_g, then so too must κ_g be zero at all stages of the unfolding process.

Imagine a surface which spans on to a non-planar boundary. If we wish to know the shape of surface which spans to this boundary but has *minimum* surface area, then the search is for a *minimal* surface. Such a surface has the property that at every point the principal curvatures are equal and opposite.

Suppose there to be two circular wire hoops placed one above the other. The surface of revolution which has minimal area and spans to these two wire hoops is a catenoid—the surface obtained by revolving a catenary about the curve's own directrix. From the nature of surface tension, soap films are minimal surfaces when not subject to a pressure difference across the film. As an approximation, the hyperbolic paraboloid is a minimum surface, when the surface is *shallow*, namely when z is everywhere small compared with the plan dimensions of the surface.

There is one further class of surface of interest which has been discussed in some detail in Chapter 5. This is the class of *constant curvature* surfaces. These surfaces are such that at all points on the surface there is *one principal curvature* of a *constant* value k, and the other principal curvature is numerically less than k.

A.14 Summary—the geometrical quantities required for the construction of a plate theory

This chapter has been devoted to the study of surface geometry. It was pointed out at the beginning that surface geometry is important because all plates when loaded deform into the shape of some surface, albeit a *shallow* one. The properties of the resulting surface shape determine specifically all the mechanical quantities of interest in constructing a mechanics of plates.

The most important induced geometrical quantities in the deformed plate are the *curvatures* and *twists*. Hence the description and analysis of curvature and twist in this chapter is the central core of technique used in this book.

However other aspects of surface geometry are also important and have been used as and when required. These other aspects have come to the fore particularly when discussing, for example, constant curvature surfaces or finite elements. When particular examples expressed in terms of particular coordinate systems are discussed, the geometrical quantities such as curvatures and twists are computed using the methods set out in this chapter.

Author Index

Subject Index

Model of the timber floor system proposed by the architect Sebastiano Serlio and first illustrated in his treatise of *c.* 1543. This view shows the arrangement of the members when supporting some load.